STUDY GUIDE TO
Chemistry and Life

An Introduction to General, Organic, and Biological Chemistry

John W. Hill
University of Wisconsin
River Falls, Wisconsin

Dorothy M. Feigl
St. Mary's College
Notre Dame, Indiana

Burgess Publishing Company
Minneapolis, Minnesota

Copyright © 1978 by Burgess Publishing Company
Printed in the United States of America
ISBN: 0-8087-3107-6

All rights reserved. No part of this book may be reproduced in any
form whatsoever, by photograph or mimeograph or by any other means,
by broadcast or transmission, by translation into any kind of language,
nor by recording electronically or otherwise, without permission in
writing from the publisher, except by a reviewer, who may quote brief
passages in critical articles and reviews.

0 9 8 7 6 5 4 3 2 1

Periodic Table of the Elements

IA	IIA		IIIB	IVB	VB	VIB	VIIB		VIII		IB	IIB	IIIA	IVA	VA	VIA	VIIA	Noble gases
1 H 1.0																		2 He 4.0
3 Li 6.9	4 Be 9.0												5 B 10.8	6 C 12.0	7 N 14.0	8 O 16.0	9 F 19.0	10 Ne 20.2
11 Na 23.0	12 Mg 24.3												13 Al 27.0	14 Si 28.1	15 P 31.0	16 S 32.1	17 Cl 35.5	18 Ar 39.9
19 K 39.1	20 Ca 40.1		21 Sc 45.0	22 Ti 47.9	23 V 50.9	24 Cr 52.0	25 Mn 54.9	26 Fe 55.8	27 Co 58.9	28 Ni 58.7	29 Cu 63.5	30 Zn 65.4	31 Ga 69.7	32 Ge 72.6	33 As 74.9	34 Se 79.0	35 Br 79.9	36 Kr 83.8
37 Rb 85.5	38 Sr 87.6		39 Y 88.9	40 Zr 91.2	41 Nb 92.9	42 Mo 95.9	43 Tc (98) ‡	44 Ru 101.1	45 Rh 102.9	46 Pd 106.4	47 Ag 107.9	48 Cd 112.4	49 In 114.8	50 Sn 118.7	51 Sb 121.8	52 Te 127.6	53 I 126.9	54 Xe 131.3
55 Cs 132.9	56 Ba 137.3		57 La* 138.9	72 Hf 178.5	73 Ta 180.9	74 W 183.9	75 Re 186.2	76 Os 190.2	77 Ir 192.2	78 Pt 195.1	79 Au 197.0	80 Hg 200.6	81 Tl 204.4	82 Pb 207.2	83 Bi 209.0	84 Po (210)	85 At (210)	86 Rn (222)
87 Fr (223)	88 Ra (226)		89 Ac† (227)	104 —§ (260?)	105 —§ (260?)	106						112?						

Key:
6 — Atomic number
C — Symbol
12.0 — Atomic mass

*
| 58
Ce
140.1 | 59
Pr
140.9 | 60
Nd
144.2 | 61
Pm
(147) | 62
Sm
150.4 | 63
Eu
152.0 | 64
Gd
157.3 | 65
Tb
158.9 | 66
Dy
162.5 | 67
Ho
164.9 | 68
Er
167.3 | 69
Tm
168.9 | 70
Yb
173.0 | 71
Lu
175.0 |

†
| 90
Th
232.0 | 91
Pa
(231) | 92
U
238.0 | 93
Np
(237) | 94
Pu
(242) | 95
Am
(243) | 96
Cm
(247) | 97
Bk
(247) | 98
Cf
(249) | 99
Es
(254) | 100
Fm
(253) | 101
Md
(256) | 102
No
(254) | 103
Lr
(256) |

‡ Parentheses around atomic mass indicate that mass given is that of the most stable known isotope.

§ Both Russian and American scientists have claimed the discovery of elements 104 and 105. Official names have not been adopted yet.

Table of Atomic Masses (Based on Carbon-12)

Element	Symbol	Atomic Number	Atomic Mass	Element	Symbol	Atomic Number	Atomic Mass
Actinium	Ac	89	(227)*	Mercury	Hg	80	200.6
Aluminum	Al	13	27.0	Molybdenum	Mo	42	95.9
Americium	Am	95	(243)	Neodymium	Nd	60	144.2
Antimony	Sb	51	121.8	Neon	Ne	10	20.2
Argon	Ar	18	39.9	Neptunium	Np	93	237.0
Arsenic	As	33	74.9	Nickel	Ni	28	58.7
Astatine	At	85	(210)	Niobium	Nb	41	92.9
Barium	Ba	56	137.3	Nitrogen	N	7	14.0
Berkelium	Bk	97	(245)	Nobelium	No	102	(254)
Beryllium	Be	4	9.01	Osmium	Os	76	190.2
Bismuth	Bi	83	209.0	Oxygen	O	8	16.0
Boron	B	5	10.8	Palladium	Pd	46	106.4
Bromine	Br	35	79.9	Phosphorus	P	15	31.0
Cadmium	Cd	48	112.4	Platinum	Pt	78	195.1
Calcium	Ca	20	40.1	Plutonium	Pu	94	(242)
Californium	Cf	98	(251)	Polonium	Po	84	(210)
Carbon	C	6	12.0	Potassium	K	19	39.1
Cerium	Ce	58	140.1	Praseodymium	Pr	59	140.9
Cesium	Cs	55	132.9	Promethium	Pm	61	(145)
Chlorine	Cl	17	35.5	Protactinium	Pa	91	231.0
Chromium	Cr	24	52.0	Radium	Ra	88	226.0
Cobalt	Co	27	58.9	Radon	Rn	86	(222)
Copper	Cu	29	63.5	Rhenium	Re	75	186.2
Curium	Cm	96	(245)	Rhodium	Rh	45	102.9
Dysprosium	Dy	66	162.5	Rubidium	Rb	37	85.5
Einsteinium	Es	99	(254)	Ruthenium	Ru	44	101.1
Erbium	Er	68	167.3	Samarium	Sm	62	150.4
Europium	Eu	63	152.0	Scandium	Sc	21	45.0
Fermium	Fm	100	(254)	Selenium	Se	34	79.0
Fluorine	F	9	19.0	Silicon	Si	14	28.1
Francium	Fr	87	(223)	Silver	Ag	47	107.9
Gadolinium	Gd	64	157.3	Sodium	Na	11	23.0
Gallium	Ga	31	69.7	Strontium	Sr	38	87.6
Germanium	Ge	32	72.6	Sulfur	S	16	32.1
Gold	Au	79	197.0	Tantalum	Ta	73	180.9
Hafnium	Hf	72	178.5	Technetium	Tc	43	98.9
Helium	He	2	4.00	Tellurium	Te	52	127.6
Holmium	Ho	67	164.9	Terbium	Tb	65	158.9
Hydrogen	H	1	1.008	Thallium	Tl	81	204.4
Indium	In	49	114.8	Thorium	Th	90	232.0
Iodine	I	53	126.9	Thulium	Tm	69	168.9
Iridium	Ir	77	192.2	Tin	Sn	50	118.7
Iron	Fe	26	55.8	Titanium	Ti	22	47.9
Krypton	Kr	36	83.8	Tungsten	W	74	183.8
Lanthanum	La	57	138.9	Uranium	U	92	238.0
Lawrencium	Lr	103	(257)	Vanadium	V	23	50.9
Lead	Pb	82	207.2	Xenon	Xe	54	131.3
Lithium	Li	3	6.94	Ytterbium	Yb	70	173.0
Lutetium	Lu	71	175.0	Yttrium	Y	39	88.9
Magnesium	Mg	12	24.3	Zinc	Zn	30	65.4
Manganese	Mn	25	54.9	Zirconium	Zr	40	91.2
Mendelevium	Md	101	(256)				

*Parentheses around atomic mass indicate that mass given is that of the most stable known isotope.

To the Student

You have a teacher and a text. Both will help you learn chemistry, but your success in the course will most likely depend on what <u>you</u> do. Listen to your teacher. Read and study the text. Study conscientiously and often. To help you organize your studying, we have provided two study aids--the problems following each chapter in the text and this study guide.

When you've completed a chapter of the text, go through the questions and exercises at the end of the chapter. These sets of problems have been designed as chapter reviews. By working through them, you can pinpoint areas of weakness in your understanding. Answers to selected problems are provided in appendix D of the text. A complete set of answers is given in this study guide. It is generally best to refer to these answers after you've tried to formulate your own. Simply reading our answers can easily lull you into a false sense of security. You read the given answer and say to yourself, "Sure, that's what I would have said." But is it? You are far more likely to understand and remember answers you've developed yourself. So use ours only to make sure you're on the right track.

This study guide is divided into sections which correspond to the chapters of the text. In these sections study hints and approaches to organizing material are given, additional practice exercises are sometimes provided, and some difficult concepts are reviewed. The study guide also contains a self-test for each chapter. When you feel you've mastered the material, take the self-test. It will provide a final check on your understanding. If you still have difficulty, seek help from your instructor.

We hope you find this study guide helpful. We invite your suggestions and criticisms. We especially solicit suggestions for improvement.

John W. Hill
Dorothy M. Feigl

Contents

Part I. Study hints, practice exercises, and self-tests

Chapter

1	Matter and Measurement	1
2	Atoms	4
3	Nuclear Processes	8
4	Chemical Bonds	11
5	Energy and Equilibria	16
6	Gases	21
7	Liquids and Solids	24
8	Oxidation and Reduction	28
9	Solutions	31
10	Acids and Bases	37
11	More Acids and Bases	41
12	Electrolytes	45
13	Bioinorganic Chemistry	50
14	Hydrocarbons	53
15	Halogenated Hydrocarbons	57
16	Alcohols, Phenols, and Ethers	60
17	Aldehydes and Ketones	64
18	Organic Acids and Derivatives	69
19	Amines and Derivatives	75
20	Compounds of Sulfur and Phosphorus	80
21	Polymers	83
22	Carbohydrates	87
23	Lipids	91
24	Amino Acids and Proteins	95
25	Nucleic Acids	99
26	Enzymes and Coenzymes	103
27	Vitamins and Hormones	107
28	Body Fluids	110
29	Digestion	114
30	Carbohydrate Metabolism	117
31	Lipid Metabolism	121
32	Protein Metabolism	125

Part II. Answers to problems in the text--Chapters 1-32   129

Part 1 Study Hints, Practice Exercises, and Self Tests
Chapter 1 Matter and Measurement

Much of the first chapter in the text is intended to place chemistry in both historical and contemporary perspective--to give you a feeling for chemistry as it affects society. If, after reading the chapter, you recognize chemistry as something more than just a course required for your particular academic program, then you have indeed understood what we were trying to say.

In addition to this overview, chapter 1 also introduces several concepts important to our further study of chemistry. These include the international system of measurement; the meaning of terms such as matter, force, and energy; different temperature scales; and so on. The problems at the end of the chapter are meant to check your understanding of this material. The following questions offer another opportunity for you to test yourself on chapter 1.

The first set of problems provides additional practice in converting among British, SI, metric, and apothecary units. You may need to refer to <u>table A.5</u> in appendix A of the text for British-metric conversion factors.

<u>Problems</u>
1. How wide in inches is 35-mm film?
2. We may define a premature baby as one which weighs less than 5 lbs at birth. What is the corresponding birth weight in the metric system?
3. If a prescription calls for 400 mg of Edrisal and each Edrisal tablet contains 6 grains, how many tablets should be taken? (There are 15 grains per gram.)
4. You are to administer 0.2 g of a medication by injection. The container label reads 100 mg/ml. How many cubic centimetres should you administer?
5. Seven grams of aspirin is enough to fatally poison a small child. If aspirin tablets contain 5 grains per tablet, how many tablets are there in a fatal dose?

<u>Multiple Choice Questions</u>

Note that many of these problems are more than tests of your memory. A number of them require preliminary calculations before an answer can be selected. You are expected to know the metric prefixes and units of measure, but you may consult a table of British-metric conversion factors. Select the <u>best</u> answer.

1. Which of the following abbreviations stands for a unit of length?
   ml   mg   dm   cc
2. The prefix <u>micro</u> is equivalent to
   10^2 10^{-2} 10^3 10^{-3} 10^6 10^{-6} 10^9 10^{-9}
3. Which of the following units of measure is equivalent to a cc?
   ml   cm   mm   mg   gr
4. How long is 1 cm?
   0.01 mm   0.2 mm   1 mm   10 mm   100 mm   1000 mm
5. How many millimetres are there in 10 cm?
   1   10   100   1000   10,000
6. How many cubic centimetres are there in a decilitre?
   0.001   0.01   0.1   1   10   100   1000
7. An object which weighs 100 μg also weighs
   0.001 mg   0.01 mg   0.1 mg   1 mg   10 mg   100 mg   1000 mg 10^{-6} g
8. If a container holds 5 ml, it will hold
   5000 ℓ   0.05 ℓ   5 cm^3 0.5 cc
9. Approximately how long in inches is a 100-mm cigarette?
   1   2   3   4   5   6
10. Lorraine is 150 cm tall and weighs 82 kg. She is
    skinny   just about perfect   a bit chubby   obese
11. If 2 ml of A weighs 4 g, then the density of A is
    0.5 g/ml   1 g/ml   2 g/ml   4 g/ml   8 g/ml

12. If a container which can hold 5 g of water is filled with 10 g of another liquid, what is the density of the liquid?
    0.5 g/ml   2 g/ml   10 g/ml   0.5   2   10   50
13. What is the specific gravity of a compound if 2 ml of it weighs 6 g?
    0.3   0.3 g/ml   3   3 g/ml   6   6 g/ml   12   12 g/ml
14. If a bottle will hold 10 g of water or 30 g of bromine, the specific gravity of bromine is
    0.33 g/ml   0.33   3 g/ml   3   30 g/ml   30   100   300
15. If a litre of a substance weighs 500 g, the specific gravity of the substance is
    500   500 g/cc   50   5   2   2 g/ml   0.5   0.05
16. If the density of a substance is 8 g/ml, what volume would 40 g of the substance occupy?
    0.2 ml   0.32 ml   0.5 ml   2 ml   3.2 ml   5 ml   20 ml   32 ml   50 ml
17. Will a bar of soap with a volume of 250 cc and a mass of 300 g float on water?
    yes   no
18. The boiling point of water is
    100 $^\circ$C   212 $^\circ$F   373 K   all of these   none of these
19. On the absolute temperature scale, temperatures are reported in
    degrees Celsius   degrees Centigrade   degrees Fahrenheit   Kelvin
20. A temperature of 98.6 $^\circ$F is the same as
    0 $^\circ$C   32 $^\circ$C   37 $^\circ$C   100 $^\circ$C   212 $^\circ$C
21. A temperature of 45 $^\circ$C is equivalent to
    -7 $^\circ$F   25 $^\circ$F   40 $^\circ$F   57 $^\circ$F   81 $^\circ$F   113 $^\circ$F   318 $^\circ$F
22. A temperature of 273 $^\circ$C is equal to
    Zero K   100 K   273 K   546 K   -273 K
23. A temperature of 77 $^\circ$F is equal to how many degrees Celsius?
    11   25   61   75   81   107   171   196   350
24. One food Calorie equals
    1000 cal   1000 kcal   1 cal   0.001 kcal
25. A cola which contains 120 Calories also contains
    0.120 cal   0.120 kcal   1200 cal   1.2 kcal   120 kcal
26. We wish to heat 10 g of water from 25 $^\circ$C to 35 $^\circ$C. The amount of heat energy required is
    0.1 cal   1 cal   10 cal   100 cal   250 cal   350 cal   1000 cal
27. If 40 cal of energy are added to 10 g of water originally at 50 $^\circ$C, the final temperature of the water will be
    unchanged   4 $^\circ$C   10 $^\circ$C   40 $^\circ$C   46 $^\circ$C   54 $^\circ$C   60 $^\circ$C   90 $^\circ$C
28. How many calories are released as 5 g of water cools from 70 $^\circ$C to 60 $^\circ$C?
    2   5   10   50   60   70   300   350
29. Which has the greatest potential energy?
    a. a small rock moving at high speed at sea level
    b. a large rock moving at slow speed at sea level
    c. a large rock balanced at the edge of a mountain top
    d. a small rock balanced on a ledge located halfway down the same mountain
    e. none of the choices is clearly the best
30. Which of the items mentioned in question 29 has the greatest kinetic energy?
    a   b   c   d   e
31. This form of matter is characterized by its tendency to maintain its volume but not its shape.
    gas   liquid   solid
32. High compressibility is a property associated with
    gases   liquids   solids
33. If two charged particles attract one another
    a. one particle must be negatively charged
    b. both particles must be negatively charged
    c. both particles must be positively charged

True or False

T F 1. Chemistry may be defined as the study of matter and the changes it undergoes.
T F 2. Chemistry is not concerned with changes in energy.
T F 3. The United States is a leader among nations changing to metric units of measure.

T F 4. Manufactured chemical products have greatly affected the American life-style.
T F 5. The ultimate source of nearly all the energy on earth is the sun.
T F 6. If measured on the moon, your mass would be different than when measured on the earth, although your weight would be the same in both places.

ANSWERS

Problems: 1) 1.4 inches, 2) 2.3 kg, 3) 1 tablet, 4) 2 cc, 5) 21 tablets

Multiple Choice:
1) dm
2) 10^{-6}
3) ml
4) 10 mm
5) 100
6) 100
7) 0.1 mg
8) 5 cm^3
9) 4
10) obese
11) 2 g/ml
12) 2 g/ml
13) 3
14) 3
15) 0.5
16) 5 ml
17) no
18) all of these
19) Kelvin
20) 37 °C
21) 113 °F
22) 546 K
23) 25
24) 1000 cal
25) 120 kcal
26) 100 cal
27) 54 °C
28) 50
29) c
30) e
31) liquid
32) gases
33) a

True or False:
1) T
2) F
3) F
4) T
5) T
6) F

Chapter 2 Atoms

A number of terms introduced in chapter 2 are basic to the language of chemistry. Because these terms are so important, we've gathered them here for easy reference.

Element--one of the basic substances from which all material things are made; the building blocks of the universe; specifically, the slightly more than 100 substances listed in the periodic table

Atom--the smallest subdivision of an element; in the modern view, a particle consisting of a massive, positively-charged core of matter (the nucleus) surrounded by a system of negatively-charged particles called electrons; atoms of an element are defined by their atomic number, i.e., the number of protons located in the nucleus of the atom

Nucleus--the highly-condensed, massive core of an atom consisting of positively-charged particles called protons and, usually, neutral particles called neutrons

Protons, Neutrons, Electrons--subatomic particles whose properties are defined in table 2.4 in the text

Atomic number--the number of protons in the nucleus of an atom; also equal to the number of electrons in a neutral atom

Atomic weight--we are using atomic weight and mass interchangeably at the moment and, therefore, the atomic weight in atomic mass units is equal to the sum of the protons and neutrons in an atom

Periodic table--an organization of the elements in order of atomic number which groups elements of similar chemical properties together; developed by Mendeleev

Ion--an atom which has gained or lost one or more electrons; a positively or negatively charged atom

Anion--negatively charged ion; an atom which has gained an extra electron (or two or three, etc.)

Cation--positively charged ion; an atom which has lost one of its electrons (or two or three, etc.)

Electronic configuration--the arrangement of electrons among energy levels in an atom

Ground state--the condition of an atom in which all electrons are in their lowest possible energy levels

Excited state--the condition of an atom in which one or more electrons occupy energy levels which are not the lowest available

Compound--a substance which incorporates two or more elements in definite (or fixed) proportions; analysis of any sample of a given compound always gives the same per cent composition

Molecule--a fixed combination of at least two atoms; all molecules of the compound water, for example, incorporate two hydrogen atoms and one oxygen atom; all molecules of the element bromine incorporate two bromine atoms

The terms associated with radioactivity (alpha, beta and gamma rays, etc.) will be reviewed with chapter 3, in which this topic is covered in far greater detail.

A number of individuals were introduced in this chapter. It will be our habit to associate developments in chemistry with the people who were responsible for them. The emphasis which individual instructors place on such biographical information will vary greatly. However, the contributions of a number of individuals mentioned in this chapter are so fundamental to the development of chemistry as a science that their names should be familiar to any student of chemistry. A selection of the most prominent individuals mentioned in this chapter is given below. See if you can identify each by associating the name with some basic chemical concept. Note that more than one concept may be associated with a single name.

Name of Person		Concept
_____Becquerel	1.	proponent of the atomic structure of matter among the ancient Greeks
_____Bohr	2.	proposed the modern atomic theory
_____Chadwick	3.	law of conservation of mass
_____Curie	4.	law of constant composition
_____Dalton	5.	law of definite proportions
_____Democritus	6.	Law of multiple proportions
_____Lavoisier	7.	made quantitative measurements the standard of chemical experimentation
_____Mendeleev	8.	the raisin pudding atom
_____Proust	9.	the nuclear atom
_____Roentgen	10.	the atom as a miniature solar system
_____Rutherford	11.	described atomic structure with complex mathematical equations
_____Schrödinger	12.	discovered the electron
_____Thomson	13.	discovered the neutron
	14.	explained the line spectra of the elements
	15.	discovered x-rays
	16.	discovered radioactivity
	17.	discovered previously unknown radioactive elements
	18.	periodic table of the elements

Numerical problems are few in this chapter. Only the law of definite proportions lends itself easily to arithmetic manipulation. After reviewing the examples in the chapter (Examples 2.1 to 2.4), try the following problems.

The analysis of the compound calcium bromide shows it to consist of 20% by weight calcium and 80% by weight bromine.

1. How many grams of calcium are present in 10 g of calcium bromide?
2. How many pounds of bromine are present in a 100-lb sample of the compound?
3. A sample of the compound weighed 50 g. How many grams of calcium and how many grams of bromine are present in this sample?
4. How large a sample of the coumpound is required to contain 5 g of calcium?
5. How many grams of bromine will combine completely with 100 g of calcium to produce the compound?
6. Is an analysis which shows that a sample of calcium bromide contains 1.5 g of calcium and 6 g of bromine consistent with the law of definite proportions?

Multiple Choice Self-Test--Refer to the periodic table (inside front cover).

Select the best answer.

1. Which term, as used by scientists, best fits this definition: a statement which summarizes the data obtained from observations?
   a. law   b. model   c. theory

2. According to the law of definite proportions, if a sample of compound A contains 10 g of sulfur and 5 g of oxygen, then another sample of A which contains 5 g of sulfur must contain
   a. 10 g of oxygen   b. 5 g of oxygen   c. 2.5 g of oxygen   d. no sample of A could contain 5 g of sulfur
3. If 1 g of sulfur dioxide contains 50% sulfur by weight, then 0.5 g of sulfur dioxide will contain what percentage of sulfur by weight.
   a. 0.5%   b. 25%   c. 50%
4. You have 10 g of element A and 10 g of element B. Compound X is known to consist of 70% A and 30% B. According to the law of definite proportions, if you mix all of your A and B together to form compound X
   a. you will get 20 g of X
   b. after all possible X has formed, some of element A will be left unreacted
   c. after all possible X has formed, some of element B will be left unreacted
   d. the compound X formed in your reaction will consist of 50% A and 50% B
5. Which of the following illustrates the law of multiple proportions?
   a. A sample of a compound contains 3 g of carbon and 4 g of oxygen; a second sample of the same compound contains 6 g of carbon and 8 g of oxygen.
   b. A sample of one compound contains 6 g of carbon and 8 g of oxygen; a sample of another compound contains 6 g of carbon and 2 g of hydrogen.
   c. A sample of one compound contains 6 g of carbon and 8 g of oxygen; a sample of another compound contains 6 g of carbon and 4 g of oxygen.
6. Which of the following facts does not fit Dalton's atomic theory?
   a. All atoms of oxygen are different from all atoms of nitrogen.
   b. Atoms are not destroyed in chemical reactions.
   c. Combinations of atoms rearrange during a chemical reaction.
   d. Oxygen atoms have the same number of protons, but may have different atomic masses.
7. The anode and a cation are
   a. positively charged   b. negatively charged   c. neutral
8. When an evacuated Crookes' tube is discharged,
   a. negative electrons jump from cathode to anode
   b. negative electrons jump from anode to cathode
   c. positive electrons jump from anode to cathode
9. When a gas-filled Crookes' tube is discharged
   a. ions formed from the gas atoms move toward the anode
   b. ions formed from the gas atoms move toward the cathode
   c. the gas atoms cannot form ions
10. Rutherford's gold foil experiment offered evidence in support of the theory that
    a. an atom has a very compact nucleus
    b. the atoms of an element can have different masses
    c. gold is a radioactive element
11. Refer to the diagram in answering the questions:

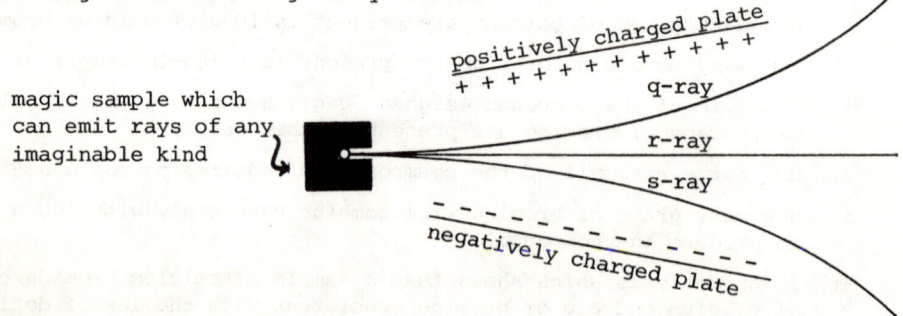

   A. The q-ray
      a. could be a stream of electrons
      b. could be a stream of protons
      c. could be either
   B. The r-ray
      a. could be a ray of something like light
      b. could be a stream of neutrons
      c. could be either

    C. The s-ray
        a. could be a stream of protons
        b. could be a stream of alpha particles
        c. could be either
12. According to the Bohr model, how many electrons are there in the highest energy level of a ground state aluminum atom?
    1   2   3   4   5   6   7   8   9   10   11   12   13   14   15   16   17   18
13. In the Bohr model of the carbon atom, the lowest energy level contains how many electrons.
    1   2   3   4   5   6   7   8   9   10   11   12   13   14   15   16   17   18
14. Which is a 2p orbital?

    a. ◯   b. ◯   c. ∞   d. ✿

15. Which element would you expect to be chemically similar to Ca?
    a. Co   b. Ba   c. K
16. A particle contains 15 protons, 14 electrons and 16 neutrons.
    a. The atomic weight of the particle is: 14   15   16   29   30   31
    b. The symbol for the element is: Si   P   S   Cu   Zn   Ga
    c. The particle has a charge of: 0   +1   +2   -1   -2
    d. If an electron were added to the particle, which of the above answers would have to be changed: a   b   c
    e. If a neutron were added to the original particle, which of the above answers would have to be changed: a   b   c
    f. If a proton were added to the original particle, which of the above answers would have to be changed: a   b   c
17. Consider the following atoms:

	Atom A	Atom B	Atom C	Atom D	Atom E	Atom F
No. of electrons	6	6	6	7	8	7
No. of neutrons	6	7	8	7	6	7
No. of protons	6	6	6	7	8	8

    a. Which atoms have the same atomic mass?   A   B   C   D   E   F
    b. For which atom(s) is the atomic number 7?   A   B   C   D   E   F
    c. Which atom(s) is(are) neutral?   A   B   C   D   E   F
    d. Which is(are) nitrogen?   A   B   C   D   E   F

ANSWERS

Scientists and their contributions:
    16 Becquerel
    10,14 Bohr
    13 Chadwick
    17 Curie
    2,6 Dalton
    1 Democritus
    3,7 Lavoisier
    18 Mendeleev
    4,5 Proust
    15 Roentgen
    9 Rutherford
    11 Schrödinger
    8,12 Thomson

Problems on definite proportions: 1) 2 g,   2) 80 lbs,   3) 10 g calcium and 40 g bromine,
    4) 25 g of calcium bromide,   5) 400 g bromine,
    6) Yes, because the analysis of the sample shows it consists of 20% calcium and 80% bromine.

Multiple Choice:
1) a   8) a   13) 2   16) e. a
2) c   9) b   14) c   f. a, b, c
3) c   10) a   15) b   17) a. C, D, E
4) c   11) A. a   16) a. 31   b. D
5) c   B. c   b. P   c. A, B, C, D, E
6) d   C. c   c. +1   d. D
7) a   12) 3   d. c

Chapter 3 Nuclear Processes

Before attempting the problems below, you should review the definitions asked for in question 1 at the end of chapter 3. It is always a good idea to do the problems at the end of the chapter in the text before attempting these exercises.

The first set of problems below is provided to give you a chance to practice balancing nuclear equations.

Complete the following equations by supplying the missing component.

1) $^{14}_{7}N + \underline{\;?\;} \longrightarrow ^{18}_{9}F$

2) $^{234}_{90}Th \longrightarrow \underline{\;?\;} + ^{0}_{-1}e$

3) $^{222}_{86}Rn \longrightarrow ^{4}_{2}He + \underline{\;?\;}$

4) $^{210}_{83}Bi \longrightarrow \underline{\;?\;} + \beta^{-}$

5) $^{56}_{26}Fe + ^{2}_{1}H \longrightarrow ^{54}_{25}Mn + \underline{\;?\;}$

6) $^{239}_{94}Pu + ^{1}_{0}n \longrightarrow \underline{\;?\;} + ^{0}_{-1}e$

7) $\underline{\;?\;} + ^{1}_{1}H \longrightarrow ^{4}_{2}He + ^{4}_{2}He$

8) $^{3}_{1}H + ^{2}_{1}H \longrightarrow ^{4}_{2}He + \underline{\;?\;}$

9) $^{241}_{95}Am + \underline{\;?\;} \longrightarrow ^{243}_{97}Bk + 2\,^{1}_{0}n$

10) $^{235}_{92}U + ^{1}_{0}n \longrightarrow ^{103}_{42}Mo + \underline{\;?\;} + 2\,^{1}_{0}n$

11) $\underline{\;?\;} + ^{1}_{0}n \longrightarrow ^{141}_{56}Ba + ^{91}_{36}Kr + 3\,^{1}_{0}n$

12) Write an equation for the alpha decay of thorium-228.

13) Polonium-218 can undergo radioactive decay by two routes. Show the product for each step.

$^{218}_{84}Po$ —beta decay→ $\underline{\;?\;}$ —alpha decay→ $\underline{\;?\;}$

$^{218}_{84}Po$ —alpha decay→ $\underline{\;?\;}$ —beta decay→ $\underline{\;?\;}$

The next set of problems provides practice in the numerical calculations involving half-lives of radioactive elements.

1) Protactinium-234 has a half-life of one minute. After 4 minutes how many micrograms of this isotope remain in a sample wiich originally contained 400 µg?

2) If the half-life of polonium-218 is 3 minutes and a sample originally contains 64 mg of this isotope, how much of the isotope remains after 15 minutes?

3) Radioactive $^{154}_{69}Tm$ has a half-life of 5 seconds. After 10 seconds, how many milligrams of this isotope remain in a sample which originally contained 160 mg?

4) Radioactive nitrogen-13 has a half-life of 10 minutes. After an hour, how much of this isotope would remain in a sample which originally contained 96 mg?

Multiple Choice Self-Test--Refer to the periodic table (inside front cover).

1. Isotopes have the same
   a. number of neutrons   b. atomic number   c. atomic weight
2. Which form of nuclear radiation most closely resembles x-rays?
   a. alpha rays   b. beta rays   c. gamma rays
3. Which is the least penetrating radiation?
   a. alpha rays   b. beta rays   c. gamma rays
4. Which is the most penetrating radiation?
   a. γ rays   b. α rays   c. β rays
5. In general, which type of radiation is most useful when diagnostic scanning of an internal organ is desired?
   a. alpha rays   b. beta rays   c. gamma rays   d. cosmic rays
6. Which of the radiation-detecting devices signals when the radiation ionizes gas molecules and causes an electric current to flow?
   a. Geiger counter   b. scintillation counter   c. film badge
7. Which type of radiation is not a stream of charged particles?
   a. alpha rays   b. beta rays   c. gamma rays   d. cosmic rays
8. Which is considered an ionizing radiation?
   a. alpha rays   b. beta rays   c. gamma rays   d. cosmic rays   e. all of these
9. Which type of radiation is stopped by a sheet of paper?
   a. alpha rays   b. beta rays   c. gamma rays
10. If the intensity of radiation is 4 units at a distance of 2 m from the source, then the intensity at 4 m from the source is
    a. 1 unit   b. 2 units   c. 4 units   d. 8 units   e. 16 units
11. Which describes the activity of the radioactive source?
    a. curie   b. roentgen   c. rad   d. rem   e. LD_{50}/30 days
12. Which is used to measure the dose of radiation absorbed by tissue?
    a. curie   b. roentgen   c. rad
13. A minimum lethal whole body dose of radiation for most human beings would be in the range
    a. 5-10 rads   b. 50-100 rads   c. 500-1000 rads   d. 5000-10,000 rads
14. You are exposed to ionizing radiation
    a. because of atmospheric testing of nuclear weapons
    b. when your teeth are x-rayed by a dentist
    c. simply because you live on the earth
    d. all of these reasons
    e. none of these reasons
15. Which reaction is currently in commercial use, providing energy for nuclear reactors?
    a. fission   b. fusion   c. radioactivity
16. Which releases more energy?
    a. fission reaction   b. fusion reaction
17. Which hydrogen isotope contains one neutron?
    a. protium   b. deuterium   c. tritium
18. Which is an isotope of $^{12}_{6}C$?
    a. $^{13}_{6}C$ b. $^{12}_{7}C$ c. $^{13}_{7}N$ d. $^{12}_{6}N$
19. For which of the following would N be an improper symbol?
    a. $^{14}_{8}X$ b. $^{14}_{7}X$ c. $^{15}_{7}X$
20. If $^{46}_{21}Sc$ emits a beta particle, the product is an isotope of
    Cl   Ar   K   Ca   Sc   Ti   V   Cr   Mn   Mo   Tc   Ru   Rh   Pd   Ag   Cd   In   Sn
21. Which isotope is particularly useful for both diagnostic and therapeutic work with the thyroid gland?
    a. cobalt-60   b. iodine-131   c. technetium-99^m
22. The isotope which has nearly ideal properties for a large number of diagnostic scanning uses, including brain scans, is:
    a. I-131   b. Tc-99^m c. U-235   d. U-238   e. Co-60

23. The sun derives its energy from
    a. fusion   b. fission   c. radioactivity
24. The difference in binding energy between reactants and products is the source of the energy released in
    a. fusion reactions   b. fission reactions   c. both types of nuclear reactions
25. Which process does this equation illustrate: $_1^2H + {}_1^2H \longrightarrow {}_2^4He$?
    a. fission   b. fusion   c. radioactivity
26. Which of the following properties makes technetium-99^m a good isotope for diagnostic scanning procedures?
    a. It does not emit alpha or beta rays.
    b. It emits extremely high-energy gamma rays.
    c. It has a very long half-life.
    d. All of these reasons.
27. Which of the following does not represent a problem associated with nuclear fission reactors used as energy sources?
    a. limited supply of fuel
    b. likelihood of an atomic-bomb-like explosion
    c. disposal of exhausted fuel
28. Which of the following nuclear events would not be described as an example of transmutation?
    a. emission of an alpha particle
    b. emission of a beta particle
    c. emission of a neutron
29. A fusion reactor is not presently used as an energy source because
    a. a source of fuel has not been located yet
    b. the chemistry of the fusion process has not be worked out
    c. the required temperature for initiating the fusion reaction is not known
    d. a procedure for achieving the necessary conditions in a controlled manner has not been worked out yet
30. A gene which has undergone mutation due to exposure to radiation can be inherited by a child if
    a. the mother was originally exposed
    b. the father was originally exposed
    c. either parent was originally exposed
    d. both parents were exposed only

ANSWERS

Nuclear equations: 1) $_2^4He$, 2) $_{91}^{234}Pa$, 3) $_{84}^{218}Po$, 4) $_{84}^{210}Po$, 5) $_2^4He$, 6) $_{95}^{240}Am$, 7) $_3^7Li$
8) $_0^1n$, 9) $_2^4He$, 10) $_{50}^{131}Sn$, 11) $_{92}^{235}U$, 12) $_{90}^{228}Th \longrightarrow {}_2^4He + {}_{88}^{224}Ra$
13) $_{85}^{218}At$ $_{83}^{214}Bi$
    $_{82}^{214}Pb$ $_{83}^{214}Bi$

Half-lives: 1) 25 µg, 2) 1 mg, 3) 40 mg, 4) 1.5 mg

Multiple Choice:
1) b   16) b
2) c   17) b
3) a   18) a
4) a   19) a
5) c   20) Ti
6) a   21) b
7) c   22) b
8) e   23) a
9) a   24) c
10) a   25) b
11) a   26) a
12) c   27) a
13) c   28) c
14) d   29) d
15) a   30) c

Chapter 4 Chemical Bonds

Chapter 4 marks a turning point in our study of chemistry. If, instead of chemistry, English literature was the object of our study, we would just be at the point of having learned to read. You can't appreciate fine literature (or even not-so-fine literature) unless you understand that those little squiggles on paper are letters of the alphabet, that letters of the alphabet are symbols which represent sounds we make to communicate with one another, that the right combinations of letters make words which have meaning.

Through the first three chapters we've been learning to use chemical symbols to represent bits of matter. In chapter 4, we've finally begun to read and write the "words" of chemistry--the formulas for compounds. In chapter 5 we'll be composing "sentences", i.e., equations, and eventually we'll be able to read and understand some of the most complex "literature" of chemistry, including the chemical version of the story of life.

Right now, it's necessary that you become comfortable in dealing with the structure of compounds. The fact that sodium chloride is an ionic compound is of great importance to its role in living systems. The fact that water is a polar covalent molecule is just as important. You must understand what ionic, covalent and polar mean to be able to understand what makes these structural features important.

Many of the problems at the end of the chapter are like grade school spelling drills or vocabulary tests. They simply ask you to practice over and over again drawing ions or putting together molecules or naming compounds. That practice should firmly establish the rules governing chemical structure in your mind. In case it hasn't, here is some additional help.

First of all, how do you know whether a compound is ionic or covalent. A fairly reliable rule states that ionic compounds are formed when a group IA or IIA element combines with a group VIA or VIIA element. (When using this rule, consider hydrogen a group VIIA element.) Group IIIA and VA elements sometimes form ionic compounds and sometimes do not. Group IVA elements tend to form covalent bonds. Since we're interested in general trends here, we will tend to stick to clear cut cases in our examples, that is, those which fit the simple rule stated in the second sentence of this paragraph (except when we are dealing with complex ions--see below).

If you are willing to commit the information in tables 4.2 and 4.4 in the text to memory (something we advise), then we can state another simple rule: ionic compounds are formed when positive and negative ions listed in these tables get together.

We emphasize recognizing ionic compounds because our quickie rule for recognizing covalent compounds is: if the compound isn't ionic, then its covalent.

Here's your chance to check your understanding of these rules and the discussion in the chapter on these points.

   I. Compounds are formed from the following sets of elements. Indicate whether the compounds would be ionic or covalent. (Note that you're not being asked to draw the compounds, just to evaluate their tendency to form ionic or covalent bonds.)
   1. Mg and O
   2. F and Ca
   3. Li and S
   4. Br and Cl
   5. Na and H
   6. S and H
   7. Ba and Br
   8. C and O
   9. N and O
   10. Rb and F

Let's stick with ionic species for the moment. Again, the complex ions listed in table 4.4 should simply be memorized (formula, including charge, and name). For simple ions, the periodic table can supply information necessary to determine the ionic charges. You should be able to draw electron dot symbols for any element in the A groups (IA, IIA, etc.). To do that you write the symbol for the element and surround it with dots representing the valence (outermost) electrons. The number of valence electrons is given by the group number.

    II. For practice, draw electron dot structures for atoms of these elements:

        1. barium   2. nitrogen   3. potassium   4. sulfur
        5. carbon   6. silicon   7. chlorine   8. boron
        9. xenon   10. hydrogen

Electron dot structures for ions are drawn by either adding electron dots to complete the shell or removing electron dots to empty the outermost shell. Electrons are added to elements with 5 or more valence electrons, they are subtracted from elements with 3 or less valence electrons.

    III. For the following elements, how many electron dots would you add to or remove from the electron dot symbols of the atoms to form ions?

        1. barium   2. nitrogen   3. potassium   4. sulfur
        5. chlorine   6. iodine   7. magnesium   8. aluminum

A charge is written to the upper right of the electron dot symbol for an ion. The charge is equal to the number of electrons added or removed in forming the ion from the neutral atom. The charge is positive if electrons are removed and negative if electrons are added.

    IV. Write electron dot symbols for the ions which are formed from the eight elements listed in problem III.

In forming ionic compounds, it is important to remember that charges in the compound must balance, i.e., add up to zero. That requirement determines the number of positive ions and negative ions which combine to form a particular compound. Examples 4.3-4.5 and 4.11-4.13 in the chapter and problem 4 at the end of the chapter permit you to check your understanding of this rule.

Once you understand ionic bonding, naming ionic compounds is a simple matter. Examples 4.6-4.8 and 4.14-4.16 and problem 5 in the text should be consulted. Remember-- the names of the ions listed in table 4.2 and 4.4 have to be memorized to work these problems.

Let's presume you are able to recognize which combinations of elements will <u>not</u> form ionic compounds (e.g., in part I, number 4, 6, 8 and 9). You are therefore dealing with covalent bonding. Putting together covalent molecules is much like working jigsaw puzzles. You move pieces around until everything fits. Once again, you may start with electron dot structures for the elements involved or with a valence bond parts list such as that given in table 4.3. If you use electron dot symbols, here are some rules of thumb to follow:

    a) Any single (unpaired) electron should be paired with a single electron on another atom.
    b) The objective is to give each atom an octet of electrons (except hydrogen, which is satisfied with a duet).
    c) This is a hint more than a rule: elements that can form only one bond (hydrogen and the group VIIA elements) should be fitted into the structure after any other elements have been attached to one another.

Here are three examples to supplement those given in the chapter.

The formula is H_2S. What is the electron dot structure? the valence bond structure?

    This is a snap. The parts are: H·   H·   ·S̈:

        One of S's single electrons and the electron of one of the H's join in a bond. The second single electron of S combines with the single electron of the remaining hydrogen.

        Voila!   H:S̈:
                       H

The valence bond structure of this molecule just replaces the shared pairs of electrons with a line between the atoms:

        H—S̈:   or   H—S   (unshared electrons not shown)
        | |
        H   H

Let's construct the CH$_4$S molecule.

Step 1   ·Ċ·   H·   H·   H·   H·   ·S̈:

Step 2   ·C̈:S̈:   H· H· H· H·
         { 4 bonds still to be   { 4 bonds to be formed by these atoms
           formed by this combination
           of atoms

Step 3   H   H
         H:C̈:S̈:   or   H—C—S̈:
            H H   H H

One last example: N$_2$H$_2$

Step 1   ·N̈·   ·N̈·   H·   H·

Step 2   ·N̈:N̈·   H·   H·

Step 3 (a dead end)   ·N̈:N̈:H   The nitrogen on the left still must form
                        H   two bonds and there are no other atoms
                                 left to bond with.

Step 4 (a retreat back to step 2)   ·N̈:N̈·   H·   H·

Step 5 (a different placement of the H's)   H:N̈:N̈:H   Each nitrogen still must
                                                      form one more bond, so they
                                                      bond to each other again.

Step 6   H:N̈::N̈:H   Everybody's happy!

         or   H—N̈=N̈—H

If the compounds in problem 9 at the end of the chapter haven't given you enough practice, try writing structural formulas (electron dot and valence bond) for some of the following: PCl$_3$, SiH$_4$, HCN, COH$_2$, ClPO, CNSH.

Polarity is the last of the general concepts dealing with chemical bonding introduced in this chapter. One speaks of polar bonds or polarity only for covalently bonded molecules, and, even then, not for all covalently bonded molecules. A covalent bond is polar only if it forms between elements of unequal electronegativity. A molecule is polar only if it contains polar bonds which do not cancel one another out.

The bond in Br-Br is not polar because the bonding atoms are of identical electronegativity; the bond in Br-F is polar because the two elements are of quite differenct electronegativities. Water is a polar molecule, but carbon dioxide is not, despite the fact that both molecules have bonds joining atoms of very different electronegativity. Water is "bent", but carbon dioxide is "straight."

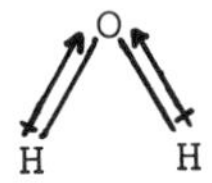

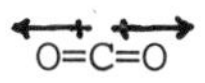

The two sets of unequally shared electrons (represented by the arrows) in carbon dioxide cancel each other out. In water, they don't. We discussed molecular shapes in this chapter because the shape of a molecule can determine whether it is polar or nonpolar.

The multiple choice self-test which follows includes a number of questions which will check your understanding of this material

Multiple Choice Self-Test: Refer to the periodic table (inside front cover).

1. The charge on an ion formed from sulfur would be

   S$^+$ S^{2+} S^{3+} S$^-$ S^{2-} S^{3-}

2. The aluminum ion is
   Al^+ Al^{2+} Al^{3+} Al^{4+} Al^{5+} Al^- Al^{2-} Al^{3-} Al^{4-} Al^{5-}
3. The ion formed from iodine is
   I^+ I^{2+} I^{3+} I^{4+} I^{5+} I^{6+} I^{7+} I^{8+} I^- I^{2-} I^{3-} I^{4-} I^{5-} I^{6-} I^{7-} I^{8-}
4. The name of the ion CN⁻ is
   a. acetate   b. ammonium   c. carbonate   d. cyanide   e. nitrate
5. How many protons are there in the Ca^{2+} ion?
   2   4   6   8   10   12   16   18   20   22   38   40   42
6. The electron dot symbol for lithium is:
   a. Li   b. Li·   c. Li·   d. ·Li·   e. ·Li·   f. ·Li: g. ·Li: h. :Li:
7. The electron dot symbol for a sulfur atom is:
   a. Ṡ   b. Ṡ·   c. Ṡ·   d. ·Ṡ·   e. ·S̈·   f. ·S̈:   g. ·S̈:   h. :S̈:
8. Which is the best formula for magnesium oxide?
   a. MgO   b. MgO_2 c. Mg_2O d. Mg_2O_2
9. Which is the correct formula for a compound of boron and sulfur?
   a. BS   b. BS_2 c. B_2S d. BS_3 e. B_3S f. B_2S_3 g. B_3S_2 h. B_3S_3
10. Which is the correct formula for a compound of aluminum and nitrogen?
    a. AlN   b. Al_3N c. AlN_3 d. Al_3N_3
11. Ferric chloride is
    a. FeCl   b. $FeCl_2$ c. $FeCl_3$ d. Fe_2Cl e. Fe_3Cl f. Fe_2Cl_3 g. Fe_3Cl_2
12. Copper (II) sulfide is
    a. CuS   b. Cu_2S c. CuS_2 d. Cu_2S_2 e. $Cu(SO_4)_2$ f. $CuSO_4$ g. Cu_2SO_4
13. Calcium nitrate is the name of
    a. $CaSO_4$ b. Ca_2SO_4 c. $Ca(CN)_2$ d. Ca_2NO_3 e. $Ca(NO_3)_2$ f. Ca_3N_2
14. Which is the most electronegative element among the following?
    a. O   b. S   c. Se   d. Te
15. Which element is most electronegative?
    a. C   b. N   c. O   d. F
16. Which is the most likely structure for a compound incorporating Be and F?
    a. Be^+F^- b. $Be^{2+}F^{2-}$ c. $Be^{2+}2F^-$ d. $2Be^+F^{2-}$ e. $Be^{2+}F^-$
17. Which would be expected to have ionic bonding?
    a. $BaCl_2$ b. NCl_3 c. $SiCl_4$
18. The compound formed from these elements would not be ionic:
    a. calcium and fluorine   b. sodium and sulfur   c. lithium and bromine
    d. sulfur and oxygen
19. Which is the best description of the bonding in ClI?
    a. Cl—I   b. Cl—I (δ+ δ-)   c. Cl—I (δ- δ+)   d. Cl^+I^- e. Cl^-I^+
20. Which compound contains a polar covalent bond?
    a. NaF   b. HF   c. F_2
21. Which of the following would have the most polar bond?
    a. F-Br   b. Cl-Cl   c. I-Br   d. F-I
22. The valence bond formula of hydrogen sulfide is
    a. H_2S b. H—S—H   c. $2H^+S^{2-}$
23. Based on bonding rules, which is not a reasonable formula?
    a. N_2 b. O_2 c. F_2 d. Ne_2
24. Which molecule is polar?
    a. Cl—Be—Cl   b. BF_3 (trigonal planar)   c. OF_2 (bent)   d. F—F

25. The CF$_4$ molecule is
    a. linear   b. bent   c. tetrahedral
26. A triple bond involves the sharing of a total of how many electrons.
    1   2   3   4   5   6   7   8
27. The correct electron dot formula for carbon dioxide is:
    a. :C:O:C: b. :O:O:C: c. :Ö::C::Ö: d. :Ö:C:Ö:
28. Which is the most reasonable valence bond formula for C$_2$F$_2$?
    a. F-C-C-F   b. C-F-F-C   c. F=C=C=F   d. F=C=C=F   e. F-C≡C-F
29. The correct electron dot formula for HNO is:
    a. H:N::Ö: b. :H:N::Ö: c. H:Ö:N: d. :H:Ö:N: e. :N::Ö:H
30. A reasonable structure for C$_2$H$_3$OCl is:

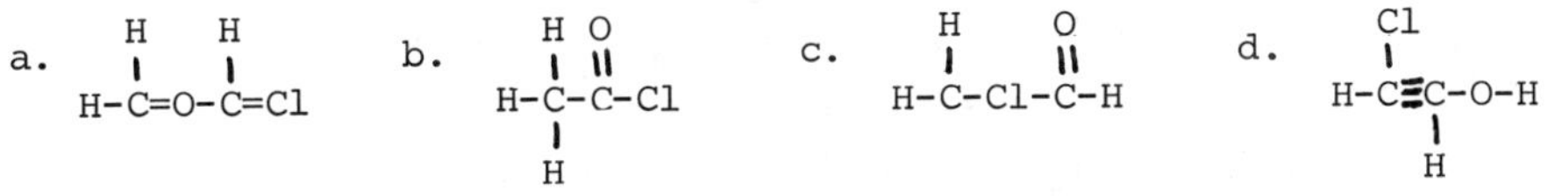

    a. H H   b. H O   c. H   O   d.   Cl
        | | | || | || |
        H-C=O-C=Cl   H-C-C-Cl   H-C-Cl-C-H   H-C≡C-O-H
                              | |
                              H   H

ANSWERS

I. Ionic: 1,2,3,5,7,10   Covalent: 4,6,8,9

II. 1) Ba·   2) ·N̈·   3) K·   4) ·S̈:   5) ·Ċ·   6) ·Ṡi·   7) ·C̈l:
    8) ·Ḃ·   9) :Ẍe: 10) H·

III. 1) remove two   2) add three   3) remove one   4) add two
     5) add one   6) add one   7) remove two   8) remove three

IV. 1) Ba^{2+} 2) N^{3-} 3) K$^+$ 4) S^{2-} 5) Cl$^-$ 6) I$^-$ 7) Mg^{2+} 8) Al^{3+}

Structural formulas: :C̈l:P:C̈l: H:S̈i:H   H:C:::N   H:C̈::Ö: :C̈l:P̈::Ö: H:N̈::C::S̈:
                          :C̈l: H   H

                       Cl-P-Cl   H-Si-H   H-C≡N   H-C=O   Cl-P=O   H-N=C=S
                          | | |
                          Cl   H   H

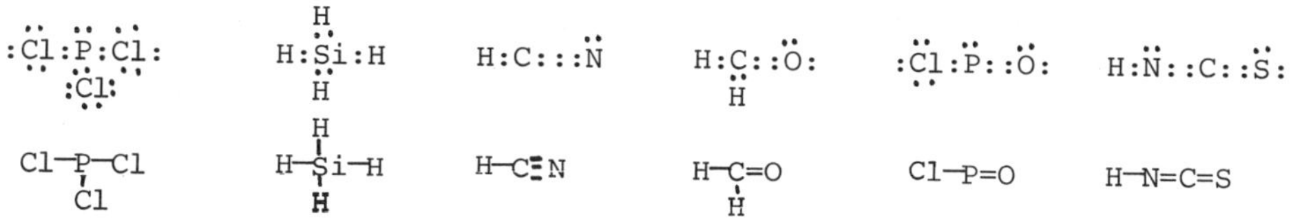

Multiple Choice:

1. S^{2-}	11. c	21. d
2. Al^{3+}	12. a	22. b
3. I$^-$	13. e	23. d
4. d	14. a	24. c
5. 20	15. d	25. c
6. a	16. c	26. 6
7. f	17. a	27. c
8. a	18. d	28. e
9. f	19. c	29. a
10. a	20. b	30. b

Chapter 5 Energy and Equilibria

The subject of this chapter is the chemical equation and the information it contains. If, having read the chapter, you find this summary of its contents too brief, then you probably never realized how much information could be pried out of an equation.

Much of the chapter is devoted to a discussion of the concepts and terminology needed if one is to extract from an equation every last bit of data. Therefore, one of the first things you should do is make sure you understand the terms collected together in question 7 at the end of the chapter.

Formula weight, mole and Avogadro's number are interrelated terms. The formula weight of a compound expressed in grams (the gram formula weight) is a mole of the compound; a mole of the compound contains Avogadro's number of the compound units, which is equal to 6.02×10^{23} units.

1 gram formula weight = 1 mole = Avogadro's number = 6.02×10^{23} units

You can treat this relationship as a multiple conversion factor. If you calculate the formula weight of a compound and remember the above relationship, then you can interconvert units expressed in grams to moles or to molecules and vice versa.

Let's examine such an interconversion involving methane, CH_4.

First we need the formula weight: atomic weight of C = 12
atomic weight of H = 1

$1 \times 12 = 12$
$4 \times 1 = \underline{4}$
16 = formula weight
16 g = gram formula weight

Now we can use that interconversion relationship. Once we have the gram formula weight we also know that there are:

16 g of CH_4/1 mole of CH_4

Avogadro's number of CH_4 molecules/16 g of CH_4

6.02×10^{23} CH_4 molecules/16 g of CH_4

How many moles are there in 4 g of CH_4?   4 g $\times \dfrac{1 \text{ mole}}{16 \text{ g}}$ = 0.25 mole

How many grams are there in 4 moles of CH_4?   4 moles $\times \dfrac{16 \text{ g}}{1 \text{ mole}}$ = 64 g

How many molecules are there in 4 g of CH_4?

4 g $\times \dfrac{6.02 \times 10^{23} \text{ molecules}}{16 \text{ g}}$ = 1.5×10^{23} molecules

How many molecules are there in 4 moles of CH_4?

4 moles $\times \dfrac{6.02 \times 10^{23} \text{ molecules}}{1 \text{ mole}}$ = 24×10^{23} molecules

Problems 2, 3 and 4 at the end of the chapter will give you practice in this area. If you'd like more, try these:

I. Calculate the formula weights. We're using relatively complicated formulas just to make sure you understand when a subscript applies to a particular atom in the formula and when it does not. You'll require a periodic table or a list of atomic weights.

1) $CaCO_3$ 2) $Be(NO_3)_2$ 3) $Al_2(C_2O_4)_3$ 4) $(CH_3)_2SO_4$

5) $(NH_4)_2CO_3$ 6) $(NH_4)_2C_2O_4$ 7) $Ca(C_2H_3O_2)_2$

II. In this problem all questions refer to the compound $C_5H_8O_2$.

1) How many moles of $C_5H_8O_2$ are there in 100 g? in 200 g? in 25 g? in 3.687 g?

2) How many grams of $C_5H_8O_2$ are there in 1 mole of the compound? in 8 moles? in 0.8 mole? in 0.01 mole?

3) How many molecules of $C_5H_8O_2$ are there in 1 mole of the compound? in 0.5 mole? in 3 moles? in 100 g? in 50 g? in 300 g?

4) a. How many carbon atoms are there in one $C_5H_8O_2$ molecule? in one mole of $C_5H_8O_2$? in 100 g of $C_5H_8O_2$?

   b. How many hydrogen atoms are there in one $C_5H_8O_2$ molecule? in one mole of $C_5H_8O_2$? in 100 g of $C_5H_8O_2$?

III. If you're still worried about your ability to interconvert these units, try answering the questions in II for the compound H_2CO_3. A calculator would be useful here (although not necessary) because the numbers won't work out so neatly. (In case you didn't notice, everything was supposed to work out rather neatly in part II.)

Let's suppose you are now totally at ease with moles and formula weights and such like. That brings us to equations. The first thing you should check in an equation is whether it is balanced or not (we'll supply the correct reactants and products). If the equation isn't balanced, the quantitative information derived from it may be incorrect. As we indicated in the chapter, you won't be balancing extremely complex equations, but you should be able to handle those shown in problem 1 at the end of the chapter. Again, for those of you who would like more practice, here are some additional equations to balance.

IV.
1) $Zn + KOH \longrightarrow K_2ZnO_2 + H_2$
2) $HF + Si \longrightarrow SiF_4 + H_2$
3) $B_2O_3 + H_2O \longrightarrow H_6B_4O_9$
4) $SiCl_4 + H_2O \longrightarrow SiO_2 + HCl$
5) $SnO_2 + C \longrightarrow Sn + CO$
6) $Fe_2O_3 + CO \longrightarrow FeO + CO_2$
7) $Fe_3O_4 + C \longrightarrow Fe + CO$
8) $Fe(OH)_3 + H_2S \longrightarrow Fe_2S_3 + H_2O$
9) $Ca_3P_2 + H_2O \longrightarrow PH_3 + Ca(OH)_2$
10) $Bi_2O_3 + C \longrightarrow Bi + CO$

Once you have a balanced equation, the coefficients in that equation give you the following information directly:
 a) the combining ratio of molecules (or other formula units like ion pairs)
 b) the combining volume ratios of gaseous reactants and products assuming temperature and pressure are maintained constant
 c) the combining ratio of moles of molecules (or other formula units)

The coefficients do not give you directly the combining weight ratios. Thus, from the equation: $CH_4 + 2\ O_2 \longrightarrow CO_2 + 2\ H_2O$

you know
 a) 1 molecule of methane (CH_4) reacts with 2 molecules of oxygen (O_2) to give 1 molecule of carbon dioxide (CO_2) and 2 molecules of water (H_2O)
 b) 1 volume of methane gas reacts with 2 volumes of oxygen gas to produce 1 volume of carbon dioxide gas and 2 volumes of water vapor
 c) 1 mole of methane reacts with 2 moles of oxygen to give 1 mole of carbon dioxide and 2 moles of water.

The equation does not say that 1 gram of methane reacts with 2 grams of oxygen to produce 1 gram of carbon dioxide and 2 grams of water. If you want to find how many grams of oxygen react with 1 gram of methane, you must first convert grams to moles and only then use the equation to determine the combining ratio. After using the equation, you'll have the answer in moles and must convert to grams.

The examples in the chapter (5.5, 5.6, 5.7, 5.14, 5.15) demonstrate the use of equations to obtain information about combining ratios. Review those examples, then problems 5 and 6 at the end of the chapter. For more practice, try some of these problems:

V. A. Refer to this equation: $CS_2 + 2\ CaO \longrightarrow CO_2 + 2\ CaS$.

1) How many moles of CO_2 are obtained from the reaction of 2 moles of CS_2? from the reaction of 2 moles of CaO?

2) How many moles of CaO are consumed if 0.3 mole of CS_2 react? if 0.3 mole of CaS are produced?

3) How many grams of CaS are obtained if 152 g of CS_2 are consumed in the reaction? if 7.6 g of CS_2 are consumed? if 22 g of CO_2 are produced? if 44 g of CO_2 are produced?

4) How many grams of CaO are required to react completely with 38 g of CS_2? with 152 g of CS_2? to produce 36 g of CaS?

B. Refer to this equation: $C_3H_8 + 5\ O_2 \longrightarrow 3\ CO_2 + 4\ H_2O + 500\ kcal$

(All compounds are gases and temperature and pressure are held constant.)

1) If 5 l of C_3H_8 react, what volume of CO_2 will be produced? What volume of O_2 will react?

2) Answer the questions in part 1 for the reaction of 5 ml of C_3H_8. for the reaction of 5 cm of C_3H_8.

3) If 22 g of C_3H_8 are burned, how many grams of CO_2 are produced? How many grams of O_2 are consumed? How many kilocalories of heat are produced?

The last question in part V.B brings energy into the discussion of chemical reactions. That question implies correctly that you treat energy like any other product or reactant, using the balanced equation to establish the ratio of energy produced (or consumed) to moles of chemical species involved. And in case you're wondering about this point, let us note that you are not able to supply the energy portion of an equation. The amount of energy required for the balanced equation would simply have to be given in a problem such as that in V.B.

Terms and concepts associated with the energy changes accompanying chemical reactions were reviewed in question 7 at the end of the chapter. We'll offer another chance to check your understanding of this terminology in the self-test below. Equilibria, le Chatliers' principle and other material treated in this chapter are also reviewed below.

Self-Test: Choose the best answer from among those offered.

1. How much does Avogadro's number of sodium atoms weigh?

   a. 11 amu   b. 23 amu   c. 11 g   d. 23 g   e. 6.02×10^{23} amu   f. 6.02×10^{23} g

2. How many atoms are there in 2 moles of helium?

   a. 2×10^{23} b. 6.02×10^{23} c. 12.04×10^{23} d. 6.02×10^{46} e. 12.04×10^{46}

3. One-half mole of SO_2 weighs

   a. 8 g   b. 12 g   c. 24 g   d. 32 g   e. 48 g   f. 64 g

4. Avogadro's number of hydrogen molecules weighs

   a. 1 g   b. 2 g   c. 3 g   d. 4 g   e. 6.02×10^{23} g

5. How many molecules of water in 6 g of H_2O?
    a. 23×10^6 b. 6×10^{20} c. 2×10^{23} d. 6×10^{23} e. 18×10^{23}
6. How many hydrogen atoms are present in 0.5 mole of H_2O?
    a. Avogadro's number   b. 0.5 x Avogadro's number   c. 2 x Avogadro's number
7. Avogadro's number is <u>not</u>
    a. the number of atoms in one mole of He
    b. the number of molecules in one mole of H_2
    c. the number of atoms in one mole of Br_2
8. If one mole of X weighs 30 g and one mole of Y weighs 15 g, then
    a. each X atom weighs twice as much as each Y atom......................T.......F
    b. 30 g of X contains twice as many atoms as 15 g of Y..................T.......F
    c. 2 moles of Y weigh twice as much as one mole of X....................T.......F
9. According to this energy diagram:

    a. Reaction I (R⟶P) is
       endothermic   exothermic
    b. For which reaction was a catalyst used?
       reaction I   reaction II
    c. For reaction I (R⟶P) the activation energy is the difference in between:
       R and T_1 T_1 and P   R and P
    d. The net energy change for reaction I (R⟶P) is the difference in energy between:
       R and T_1 T_1 and P   R and P
    e. For the reverse reaction for I (P⟶R) the net energy change for the reaction results in energy being ___?___ by the system.
       absorbed   released   there is no net change in energy
10. An increase in reaction rate is <u>not</u> expected to accompany an increase in
    a. concentration of reactants   b. temperature   c. activation energy
11. In general, which reaction would be expected to have the higher activation energy?
    a. a strongly exothermic reaction   b. a strongly endothermic reaction
12. A catalyst
    a. increases the activation energy and increases the rate of a reaction
    b. decreases the activation energy and decreases the rate of the reaction
    c. increases the activation energy and decreases the rate of the reaction
    d. decreases the activation energy and increases the rate of the reaction
13. An increase in temperature generally results in an increase in the rate of a reaction. Which of the following does <u>not</u> account, at least in part, for this phenomenon?
    a. reacting particles collide more frequently
    b. collisions of faster moving particles are more likely to supply the activation energy for the reaction
    c. at higher temperature reactions change from endothermic to exothermic
14. Refer to this reaction in which all compounds are gases: $2A + 2B \rightleftarrows 3X + Y$ + heat.
    The reaction is originally at equilibrium. Predict the direction of the shift in equilibrium for the following:
    a. addition of compound A   right   left   no change
    b. removal of compound X   right   left   no change
    c. an increase in pressure   right   left   no change
    d. an increase in temperature   right   left   no change
    e. addition of a catalyst   right   left   no change

ANSWERS

I. 1) 100   2) 133   3) 318   4) 126   5) 96   6) 124   7) 158

II. The formula weight of $C_5H_8O_2$ = 100

1) 1 mole; 2 moles; 0.25 mole; 0.03687 mole
2) 100 g; 800 g; 80 g; 1 g
3) 6.02×10^{23}; 3.01×10^{23}; 18.06×10^{23}; 6.02×10^{23}; 3.01×10^{23}; 18.06×10^{23}
4) a. 5; 5 moles or 30×10^{23}; 5 moles or 30×10^{23}
   b. 8; 8 moles or 48×10^{23}; 8 moles or 48×10^{23}

III. The formula weight of H_2CO_3 = 62

1) 1.6 moles; 3.2 moles; 0.40 mole; 0.059 mole
2) 62 g; 496 g; 49.6 g; 0.62 g
3) 6.02×10^{23}; 3.01×10^{23}; 18.06×10^{23}; 9.71×10^{23}; 4.85×10^{23}; 29.1×10^{23}
4) a. 1; 1 mole or 6×10^{23}; 1.6 moles or 9.71×10^{23}
   b. 2; 2 moles or 12×10^{23}; 3.2 moles or 19.4×10^{23}

IV. 1) $Zn + 2\ KOH \longrightarrow K_2ZnO_2 + H_2$
2) $4\ HF + Si \longrightarrow SiF_4 + 2\ H_2$
3) $2\ B_2O_3 + 3\ H_2O \longrightarrow H_6B_4O_9$
4) $SiCl_4 + 2\ H_2O \longrightarrow SiO_2 + 4\ HCl$
5) $SnO_2 + 2\ C \longrightarrow Sn + 2\ CO$
6) $Fe_2O_3 + CO \longrightarrow 2\ FeO + CO_2$
7) $Fe_3O_4 + 4\ C \longrightarrow 3\ Fe + 4\ CO$
8) $2\ Fe(OH)_3 + 3\ H_2S \longrightarrow Fe_2S_3 + 6\ H_2O$
9) $Ca_3P_2 + 6\ H_2O \longrightarrow 2\ PH_3 + 3\ Ca(OH)_2$
10) $Bi_2O_3 + 3\ C \longrightarrow 2\ Bi + 3\ CO$

V. A. 1) 2 moles of CO_2; 1 mole of CO_2
2) 0.6 mole of CaO; 0.3 mole of CaO
3) 288 g of CaS; 14.4 g of CaS; 72 g of CaS; 144 g of CaS
4) 56 g of CaO; 224 g of CaO; 28 g of CaO

B. 1) 15 l of CO_2; 25 l of O_2
2) for 5 ml of C_3H_8: 15 ml of CO_2; 25 ml of O_2
   for 5 m^3 of C_3H_8: 15 m^3 of CO_2; 25 m^3 of O_2
3) 66 g of CO_2; 80 g of O_2; 250 kcal

Self-Test

1. d
2. c
3. d
4. b
5. c
6. a
7. c
8. a. T
   b. F
   c. F
9. a. endothermic
   b. reaction II
   c. R and T_1
   d. R and P
   e. released
10. c
11. b
12. d
13. c
14. a. right
    b. right
    c. no change
    d. left
    e. no change

Chapter 6 Gases

As always, we shall assume that you've worked through the problems at the end of the chapter in the text. You should, therefore, have a pretty good idea of how well you understand the gas laws of Boyle, Charles and Gay-Lussac. The following diagram presents the briefest summary of these laws:

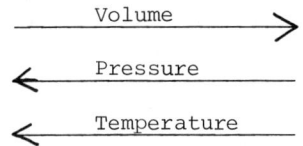

The diagram is supposed to call to mind which of the variables vary directly and which vary inversely. Thus, volume and pressure move in opposite directions, as do volume and temperature. Pressure and temperature move together. Also remember that any units of volume and pressure can be used in these gas laws as long as you are consistent, but temperature variations must be calculated on the absolute scale (in Kelvin). Beyond these few reminders, the best thing we can do to help you with these gas laws is to supply a few more practice problems to sharpen your abilities.

I. A gas with a pressure of 500 mm Hg at 0 °C occupies a volume of 400 ml.
   1) If the temperature is changed to 273 °C while the pressure is held constant, what will the new volume be?
   2) If the volume of the original gas was compressed to 20 ml and the temperature was maintained constant, what would the pressure be?
   3) What would the volume be at STP?

II. If a gas occupies 2 ℓ at 27 °C and 1 atm:
   1) what would the volume be if the temperature remained constant but the pressure decreased to 0.25 atm?
   2) what would the volume be if the pressure remained constant but the temperature decreased to -123 °C?
   3) what would the volume be at STP?

III. A balloon was filled with gas to a volume of 60 ℓ at a temperature of 30 °C and a pressure of 750 torr.
   1) What will the volume of the balloon be if it drifts to a point where the temperature is still 30 °C but the pressure is only 250 torr?
   2) What will the volume of the balloon be if the pressure remains constant but the temperature drops to -71 °C?
   3) What would the volume be at STP?

IV. A rigid container is filled with gas at 68 °F and 15 psi and then sealed. What will the pressure inside the container be if it is heated to 212 °F?

For the remainder of the material in chapter 6, we'll rely on the self-test to check your understanding. Remember that some of the self-test questions do require calculations like the problems above, that is, they are asking for more than simple recall of factual material. You may refer to a periodic table if necessary (for atomic weights, for example), and you will have to refer to table 6.2 in the text for some problems.

Self-Test

1. Which gas has the highest partial pressure in the atmosphere?
   a. CO_2 b. H_2O c. O_2 d. N_2
2. If a pressure is reported as 1 torr, it is also
   a. 760 mm Hg   b. 760 atm   c. 1 mm Hg   d. 1 atm
3. The conditions expressed in the abbreviation STP do not include
   a. 0 °C   b. 273 K   c. 760 mm Hg   d. 1 atm   e. 1 torr

4. According to the kinetic molecular theory, if you increase the temperature of a gas without changing the volume of the gas, the particles of the gas will
   a. strike the walls more often
   b. strike the walls with less force
   c. lose kinetic energy
   d. increase in size
5. Boyle's law would be explained by the kinetic molecular theory in the following way:
   a. If the volume of a gas sample decreases, the particles move faster and strike the walls of the container more often, and this is measured as an increase in pressure.
   b. If the volume of a gas is decreased, the distance between the walls of the container decreases and the particles strike the walls more often. This results in an increase in pressure.
   c. Both of the above explanations are incorrect because a decrease in volume results in a decrease in the pressure of a gas.
6. At increased temperatures, gas molecules
   a. move faster
   b. become larger
   c. attract one another
   d. lose kinetic energy
7. An increase in the pressure of a gas corresponds to
   a. an increase in the concentration of the gas
   b. a decrease in the concentration of the gas
   c. an increase in the size of the particles of the gas
8. In applying the following gas laws, which variable is held constant?
   a. Boyle's law   pressure   temperature   volume
   b. Gay-Lussac's law   pressure   temperature   volume
   c. Charles' law   pressure   temperature   volume
9. Henry's law can be used to explain the
   a. operation of an iron lung
   b. bends
   c. process of respiration
10. If a gas occupies 4 ℓ at 2 atm, how many litres will it occupy at 4 atm if the temperature isn't changed?
    a. 1   b. 2   c. 4   d. 8   e. 16   f. 32
11. If a gas occupies 1 ℓ at 273 °C, how many litres will it occupy at 0 °C if the pressure does not change?
    a. 1/273   b. 0.5   c. 1   d. 2   e. 4   f. 273
12. If a mixture of gases has a total pressure of 100 mm Hg and the partial pressure of nitrogen in the mixture is 25 mm Hg, then the per cent of nitrogen in the mixture is
    a. 4%   b. 40%   c. 400%   d. 0.25%   e. 2.5%   f. 25%   g. 125%   h. 250%
13. If the concentration of water vapor in the air is 1% and the total atmospheric pressure equals 1 atm, then the partial pressure of water vapor is
    a. 0.1 atm   b. 1 mm Hg   c. 7.6 mm Hg   d. 100 atm   e. 760 mm Hg
14. If a sample of gas at STP contains 10% helium, what is the partial pressure of helium in the sample?
    a. 0.1 atm   b. 0.1 mm Hg   c. 10 atm   d. 10 mm Hg   e. 7.6 atm   f. 7.6 mm Hg
15. If the relative humidity is 50% on a day when the temperature is 30 °C, the partial pressure of water vapor in the air is
    a. 16 mm Hg   b. 30 mm Hg   c. 50 mm Hg   d. 32 mm Hg   e. 64 mm Hg
16. Nitrogen is collected over water at 30 °C. If the total pressure within the collection vessel is 1 atm, the partial pressure of nitrogen is
    a. 0.68 atm   b. 32 torr   c. 728 torr   d. 760 mm Hg   e. 792 mm Hg
17. The molar volume of helium gas (atomic number 4) is
    a. 4 g/ℓ   b. 4 moles   c. 4 ℓ   d. 6.02×10^{23} ℓ   e. 22.4 ℓ
18. A flask contains Avogadro's number of neon atoms at STP. The volume of the flask is how many litres?
    a. 1   b. 10   c. 20   d. 22.4   e. 6.02×10^{23}
19. If a container holds 0.5 mole of N_2 gas at STP, the container holds how much N_2?
    a. $6.02 \times 10^{11.5}$ molecules   b. 0.5×10^{23} molecules   c. 7 g   d. 14 g   e. 28 g

20. If a container holds 2 g of helium at STP, how big is the container?
    a. 0.5 ℓ   b. 11.2 ℓ   c. 20 ℓ   d. 22.4 ℓ   e. 44.8 ℓ
21. The approximate density of CO_2 gas at STP is
    a. 1 g/ml   b. 2 g/ml   c. 2 g/ℓ   d. 44 g/ml   e. 44 g/ℓ   f. 22.4 g/ℓ
22. What is the approximate weight of 10 ℓ of C_3H_8 gas?
    a. 10 g   b. 20 g   c. 22.4 g   d. 44 g   e. 440 g
23. The bends result from which dissolved gas being released as a deep sea diver ascends?
    a. H_2O b. O_2 c. N_2 d. CO_2 e. He
24. The mechanism by which we inspire air depends on our ability to __?__ the size of our chest cavity.
    a. increase   b. decrease
25. In respiratory therapy, the highest humidity is imparted to a gas mixture to be administered through the
    a. mouth   b. nose   c. trachea
26. The formation of oxyhemoglobin is favored by __?__ P_{O_2}.
    a. high   b. low
27. If the partial pressure of a gas in the alveoli is 104 torr and the gas tension of venous blood is 40 torr
    a. the net diffusion of the gas will be from alveoli to blood
    b. the net diffusion of the gas will be from blood to alveoli
    c. nothing happens because gas tension and partial pressure have nothing to do with one another
28. Where would you expect the P_{CO_2} to be highest?
    a. in the alveoli immediately after inspiration of fresh air
    b. in the venous blood
    c. in the arterial blood
    d. in a cell which is metabolically active

ANSWERS
I.   1) 800 ml;   2) 10,000 mm Hg;   3) 263 ml
II. 1) 8 ℓ;   2) 1 ℓ;   3) 1.82 ℓ
III. 1) 180 ℓ;   2) 40 ℓ;   3) 53 ℓ
IV. 19 psi

Self-Test:

1. d
2. c
3. e
4. a
5. b
6. a
7. a
8. a. temperature
   b. volume
   c. pressure
9. b
10. b
11. b
12. f
13. c
14. a
15. a
16. c
17. e
18. d
19. d
20. b
21. c
22. b
23. c
24. a
25. c
26. a
27. a
28. d

Chapter 7 Liquids and Solids

Many new terms were introduced in chapter 7. A knowledge of these is essential to an understanding of the concepts introduced in the chapter. Therefore, we have gathered the terms with their definitions here for easy reference.

Dipole--a molecule which contains a center of negative and a separate center of positive charge; polar molecule; a molecule which will adopt a preferred alignment in an electric field

Hydrogen bond--an unusually strong attractive force between polar molecules which incorporate the following structural features: each of the interacting molecules must include a small, strongly electronegative element (N, F or O) and at least one of the molecules must have a hydrogen atom covalently bonded to that electronegative element

Dispersion forces--transient attractive forces between molecules caused by the interaction of momentary dipoles which result from the movement of electrons within orbitals

Viscosity--the tendency of a fluid to resist flow; when a liquid is described as "thin" or "thick", the property being described is viscosity

Surface tension--the tendency of the surface of a liquid to contract because of unbalanced intermolecular forces; this phenomenon is responsible for the surface's tendency to exhibit some of the properties of a thin, stretched membrane

Vapor/gas--We have used the terms gas and vapor almost interchangeably, and you should regard them as equivalent. They can be more precisely defined. The term gas is used if the substance referred to exists in the gaseous state at room temperature and normal atmospheric pressure. A substance which is ordinarily a liquid or solid at room temperature is called a vapor when converted to the gaseous state. Thus, the two terms do really describe the same condition of matter. If the normal boiling point of the substance happens to be below room temperature we say "gas". If the boiling point is above room temperature, we say "vapor".

Vaporization--the process in which a substance changes from the liquid to the gaseous state

Condensation--the process in which a substance changes from the gaseous to the liquid state

Distillation--the process in which a liquid is vaporized and recondensed in order to purify it

Boiling point--the temperature at which the vapor pressure of a liquid equals atmospheric pressure; the temperature at which vaporization occurs throughout the body of a liquid sample rather than at the surface only; the normal boiling point is the boiling point at a pressure of 760 mm Hg or 1 atm

Melting point--the temperature at which the transition from solid to liquid (or vice versa) takes place for a given substance; at the melting point, solid and liquid states are in equilibrium -- addition of energy (heat) will shift the equilibrium toward the liquid phase (the solid melts), removal of energy (cooling) will shift the equilibrium toward the solid phase (the liquid freezes)

Heat of vaporization--the amount of energy required to convert a stated amount of liquid to vapor; the equivalent amount of energy is transferred to the surroundings when the same amount of vapor condenses to the liquid state; the molar heat of vaporization is the energy required to convert one mole of the compound from liquid to gas or the amount of energy released in the conversion of one mole of compound from gas to liquid

Heat of fusion--the amount of energy required to convert a stated amount of solid to liquid; the amount of energy released to the surroundings when a stated amount of liquid solidifies; molar heat of fusion is the energy released when one mole of liquid solidifies or the equivalent amount of energy absorbed when one mole of the solid liquifies

Heat capacity--the amount of heat required to change the temperature of a stated amount of a substance; a high heat capacity means the substance can absorb a relatively large amount of energy before becoming hot; a low heat capacity means that the substance will get hot upon the absorption of relatively small amounts of energy

Specific heat--a term related to heat capacity, but defined in specific units, i.e., the amount of energy in calories required to raise the temperature of 1 g of a substance 1 °C

Crystal lattice--the arrangement in a solid of ions, molecules or atoms to give a relatively rigid and highly ordered pattern

Unit cell--the smallest repeating unit of a crystal lattice

Ionic solid--those solids in which ions occupy the lattice points in the crystal; sodium chloride is an example

Metallic solid--a solid in which the lattice points are occupied by metal atoms whose outer electrons are free to move like a fluid throughout the lattice

Molecular crystal--a solid in which the points of the crystal lattice are occupied by discrete molecules; ice is an example

Covalent network crystal--a solid in which the crystal lattice consists of a network of covalent bonds joining atoms at the lattice points; the lattice of such a crystal is not defined by many individual molecules, but coincides with what is essentially a single, huge covalent molecule; diamond is an example

Numerical problems in this chapter deal with energy changes associated with changes of state and changes of temperature. Energy must be supplied to a system to convert a solid to a liquid, to convert a liquid to a gas, or to raise the temperature of a gas or a liquid or a solid. The reverse changes (liquid to solid, gas to liquid, or decreases in temperature) occur if energy is removed from the system.

The system gains or loses energy in definite stages. You can raise the temperature of a solid by supplying it with energy only until you reach the melting point. Then any energy you add will be used up in converting solid to liquid. While this change is occurring, the temperature stays the same (at the melting point). As soon as the solid is completely converted to a liquid, additional energy will be used to increase the temperature of the liquid. Eventually the boiling point of the liquid is reached. Then the temperature remains constant while additional energy is used to convert liquid to gas. When this conversion is complete, the temperature of the gas can be raised by supplying more energy.

To calculate the energy needed for each of these changes, you need the following constants:

Change Effected	Constant Used
Increase or decrease temperature of the solid	Specific heat of the solid
Conversion of solid to liquid or vice versa	Heat of fusion
Increase or decrease in temperature of the liquid	Specific heat of the liquid
Conversion of liquid to gas or vice versa	Heat of vaporization
Increase or decrease in temperature of the gas	Specific heat of the gas

A change which starts with a solid and ends with a gas should simply be treated as a multistep calculation (see example 7.6 in the text). Problems at the end of the chapter offer practice in dealing with these energy changes. A few more are offered below for additional practice. Since the energy changes involving water are so important to life on this planet, all of the additional problems deal with this substance.

The pertinent constants are:

specific heat of ice = 0.5 cal/g/°C

specific heat of water = 1 cal/g/°C

heat of fusion of water = 80 cal/g

heat of vaporization of water = 540 cal/g

Problems

1. One kilogram of water at 20 °C was heated to 30 °C. How much energy was required?
2. A sample of water weighing 50 g cooled from 70 °C to 20 °C. How much heat was released to the surroundings?
3. How many calories are required to convert 15 g of water to steam at 100 °C?
4. How many calories are required to convert 15 ml of water, originally at 20 °C to steam at 100 °C?
5. Approximately how much heat is removed from the body in one day by the evaporation of 500 ml of perspiration?
6. How many calories are required to convert 2 moles of ice to water at 0 °C?
7. If 1 g of steam at 100 °C is condensed, cooled and converted to ice at 0 °C, how much energy is released to the surroundings?

Bonding - The types of bonding in column B are responsible for maintaining the crystal lattice structure of the materials in column A. Choose the type of bonding in B which plays the most significant role for each item in A.

Column A	Column B
_____ I_2	1. ionic bonding
_____ H_2O	2. dipole interactions
_____ Ne	3. hydrogen bonding
_____ Na	4. dispersion forces
_____ NaCl	5. metallic bonding
_____ ICl	

Self-Test - Select the best answer

1. At room temperature ionic compounds exist as
   a. solids   b. liquids   c. gases
2. For which compound would you expect the interionic forces to be strongest?
   a. LiF   b. BeO   c. BN
3. For which compound would the dispersion forces be greatest?
   a. Cl_2 b. Br_2 c. I_2
4. Which type of force would not be operating in a solid sample of HCl?
   a. dipolar forces   b. dispersion forces   c. interionic forces
5. Which represents the best arrangement for a pair of dipoles?
   a. (+ − + −)   b. (+ − − +)   c. (+ −) / (− +)   d. (+ −) / (+ −)
6. Which of the following correctly illustrates hydrogen bonding?
   a. H—N with H, H····O—H   b. F—H····H—O with H   c. Cl—H····Cl—H
7. For which compound is hydrogen bonding a significant attractive force?
   a. F—N=O   b. Cl—N=O   c. H—N=O   d. H—C≡C—H
8. In general, which is the strongest type of bonding?
   a. ionic bonding   b. dipole interactions   c. hydrogen bonding
9. Which is the most highly ordered arrangement of matter?
   a. gas   b. liquid   c. solid
10. Which compound would you expect to have the highest melting point?
    a. C_3H_8 b. C_8H_{18} c. $C_{18}H_{38}$

11. If compound A has a lower vapor pressure than compound B at a given temperature, which compound would be expected to have the lower boiling point?
    a. A   b. B   c. the boiling point would be the same at 760 mm Hg
12. The atmospheric pressure is 670 mm Hg. Water will boil at
    a. less than 100 °C   b. exactly 100 °C   c. more than 100 °C
13. If a liquid and its vapor are at equilibrium in a closed container and the temperature is increased, which rate increases?
    a. rate of vaporization   b. rate of condensation   c. both   d. neither
14. Which process makes your breath visible on a cold day?
    a. condensation   b. vaporization   c. fusion
15. In general, which is expected to be larger for a given substance?
    a. molar heat of fusion;   b. molar heat of vaporization;   c. specific heat
16. Which would be expected to have the highest heat of vaporization?
    a. CH_4 b. H_2O c. LiF
17. When is the viscosity of a lubricating oil highest?
    a. at room temperature   b. at 150 °C   c. at -15 °C
18. If you start with 1 g of ice at 0 °C and add 100 calories of heat energy, what will the temperature of the water be?
    0 °C   10 °C   20 °C   30 °C   40 °C   50 °C   60 °C   70 °C   80 °C   90 °C   100 °C
19. Refer to the following specific heats: For x = 0.1 cal/g/°C
    For y = 0.4 cal/g/°C
    For z = 0.8 cal/g/°C
    a. Which material has the highest heat capacity?   x   y   z
    b. If all are supplied with the same amount of heat, which will reach the highest temperature?   x   y   z
20. Which of the following represents the expected arrangement of particles in a water solution of sodium chloride?

    a. Na^+ $\delta-O$ with $\delta+H$, $\delta+H$ b. Na^+ $\delta+H$, $\delta+H$ with $O\delta-$ c. Cl^- $\delta-O$ with $H\delta+$, $H\delta+$

21. Which property of water is regarded as unusual?
    a. physical state at room temperature
    b. the relative density of solid and liquid
    c. heat of vaporization
    d. specific heat
    e. all of these
    f. none of these

ANSWERS

Problems: 1) 10 kcal; 2) 2.5 kcal; 3) 8.1 kcal; 4) 1.2 kcal; 5) 270 kcal; 6) 2.88 kcal;
           7) 0.72 kcal

Bonding: __4__ I_2
           __3__ H_2O One could argue that hydrogen bonding is simply a form of dipole interaction. However, hydrogen bonding is sufficiently distinctive to warrant a separate category.
           __4__ Ne
           __5__ Na
           __1__ NaCl
           __2,4__ ICl   It is difficult to say whether dipole interactions or dispersion forces would be more important for this compound.

Self-Test:
1. a
2. c
3. c
4. c
5. c
6. a
7. c
8. a
9. c
10. c
11. b
12. a
13. c
14. a
15. b
16. c
17. c
18. 20 °C
19. a. z; b. x
20. a
21. e

Chapter 8 Oxidation and Reduction

It is impossible to overemphasize the importance of oxidation-reduction processes. Think of it this way. You are powered by the energy of sunlight. Only you can't simply unfold solar panels as artificial satellites do and convert sunlight to stored electrical energy to be tapped as necessary. You are a chemical factory and not an artificial satellite, solar-powered or otherwise. So somehow you have to tap that solar energy in a chemical way. That is precisely the role of oxidation-reduction reactions in life processes--they plug you into the sun. The later chapters of the text will detail the sequence of reactions which accomplishes this objective. Right now we are concerned with familiarizing ourselves with the general features of oxidation-reduction reactions.

We've spent much time in this chapter developing the ability to recognize when a compound is oxidized or reduced. Included in this consideration was the concept of oxidation number. On first encounter, there is something mysterious and magical about oxidation numbers. We have no objection to their being considered magical because they are. They were conjured up by chemists to make it easier to deal with oxidation-reduction reactions. It is the mystery of oxidation numbers we have tried to dispel. They simply represent a means of assessing oxidation or reduction. You should be able to look at an equation and recognize when oxidation (and reduction) has occurred. (Always remember that one process is impossible without the other.) If you can do that without referring to oxidation numbers--fine. But we think you'll find, just as chemists already have, that occasionally a consideration of oxidation numbers makes the evaluations process a little easier.

In problems I and II below we're offering additional practice in determining oxidation numbers and in evaluating the components of oxidation-reduction equations. These problems supplement problems 2 and 3 at the end of the chapter.

I. Determine the oxidation numbers of the underlined atoms.

   a. $\underline{O}_3$
   b. $\underline{N}_2$
   c. $\underline{N}H_3$
   d. $\underline{Cr}O_3$
   e. $\underline{Cr}_2O_7^{2-}$
   f. $NaO\underline{Cl}$
   g. $H\underline{Cl}O_4$
   h. $H\underline{Cl}O_2$
   i. $K\underline{I}O_3$
   j. $H_5\underline{I}O_6$
   k. $H_3\underline{P}O_4$
   l. $H_4\underline{P}_2O_7$
   m. $Ca H\underline{P}O_4$
   n. $H\underline{S}_2O_7^-$
   o. $\underline{N}_2O_3$
   p. $\underline{N}_2O_4$
   q. $\underline{N}_2O_5$
   r. $\underline{N}O_2^-$
   s. $\underline{C}_2H_6$
   t. $\underline{C}H_2O_2$
   u. $\underline{C}_2H_3O_2$
   v. $\underline{C}H_4O$
   w. $\underline{C}O_2$
   x. $\underline{C}_3H_6O_3$
   y. $\underline{C}_3H_4O_5$
   z. $\underline{C}_{12}H_{22}O_{11}$

II. Identify the element being oxidized, the element being reduced, the oxidizing agent and the reducing agent for each equation.

   a. $4\,Na + CO_2 \longrightarrow 2\,Na_2O + C$
   b. $C_2H_4 + 3\,O_2 \longrightarrow 2\,CO_2 + 2\,H_2O$
   c. $2\,Ag^+ + Cu \longrightarrow 2\,Ag + Cu^{2+}$
   d. $5\,CO + I_2O_5 \longrightarrow I_2 + 5\,CO_2$
   e. $3\,SO_2 + 2\,CrO_3 + 3\,H_2O \longrightarrow Cr_2O_3 + 3\,H_2SO_4$

Self-Test: Select the best answer
1. The earth's crust does <u>not</u> contain a large abundance of oxygen in the form of
   a. O_2 b. H_2O c. H_2O_2 d. SiO_2
2. A substance is oxidized if it
   a. gains oxygen atoms   b. gains hydrogen atoms   c. gains electrons
   d. all of these   e. none of these

3. In the reaction, $C_2H_4 + H_2O \longrightarrow C_2H_6O$, carbon is
   a. oxidized   b. reduced   c. neither oxidized nor reduced
4. Hydrogen is no longer used in lighter-than-air ships because
   a. it is more dense than helium
   b. it is a strong oxidizing agent
   c. under certain conditions it reacts explosively to form water
   d. it is too difficult to obtain from petroleum products
5. Platinum is a useful catalyst in reactions involving hydrogen because
   a. the hydrogen absorbed on the catalyst's surface is more reactive than molecular hydrogen
   b. the activation energy of the reaction using platinum is lower
   c. both of the above statements are correct
   d. neither of the above statements is correct
6. Two common oxidizing agents change color when they are reduced. They are (select two):
   a. H_2O_2 b. $KMnO_4$ c. $Na_2Cr_2O_7$ d. O_2 e. NaOCl
7. Which is not used as an oxidizing agent?
   a. Cl_2 b. HNO_3 c. NaOCl   d. C
8. Which compound can be an oxidizing agent or a reducing agent?
   a. H_2O_2 b. $KMnO_4$ c. H_2
9. When hydrogen peroxide is used as a reducing agent, it is oxidized to
   a. O_2 b. H_2O c. H_2O_2 d. H_2
10. Disinfectants, antiseptics and bleaches are frequently
    a. oxidizing agents   b. reducing agents
11. In photosynthesis, carbon dioxide is __?__ to a sugar.
    a. oxidized   b. reduced
12. Which is not a reason that ozone is preferred to chlorine as a disinfectant for water?
    a. Ozone is less expensive than chlorine.
    b. Chlorine has been shown to form toxic byproducts.
    c. Ozone is more effective against some viruses.
    d. Chlorine imparts a distinctive taste to water.
13. In general, animals, including humans, are
    a. oxidizing agents   b. reducing agents
14. The oxidation number of Cl in $HClO_4$ is
    +1   +2   +3   +4   +5   +6   +7   +8   -1   -2   -3   -4   -5   -6   -7   -8
15. The oxidation number of Sn in SnO_3^{2-} is
    +1   +2   +3   +4   +5   +6   +7   +8   -1   -2   -3   -4   -5   -6   -7   -8
16. The oxidation number of P in $H_2PO_3^-$ is
    +1   +2   +3   +4   +5   +6   +7   +8   -1   -2   -3   -4   -5   -6   -7   -8
17. In the equation, $FeO + C \longrightarrow Fe + CO$,
    a. FeO is an(a)   oxidizing agent   reducing agent
    b. C is   oxidized   reduced
18. According to the equation, $Ca + Cu^{2+} \longrightarrow Ca^{2+} + Cu$,
    a. calcium is being   oxidized   reduced
    b. copper ion is the   oxidizing agent   reducing agent
19. In the reaction, $N_2 + 2 H_2O \longrightarrow NH_4^+ + NO_2^-$, nitrogen is:
    a. oxidized   b. reduced   c. both   d. neither
20. Consider the reaction: $3 C_2H_4 + 2 MnO_4^- + 4 H_2O \longrightarrow 3 C_2H_6O_2 + 2 MnO_2 + 2 OH^-$.
    a. What is the oxidation number of C in C_2H_4?
       +1   +2   +3   +4   +5   -1   -2   -3   -4   -5   0
    b. What is the oxidation number of C in $C_2H_6O_2$?
       +1   +2   +3   +4   +5   -1   -2   -3   -4   -5   0
    c. Is carbon being oxidized or reduced?
       Oxidized   Reduced

ANSWERS
Problem I: a) 0; b) 0; c) -3; d) +6; e) +6; f) +1; g) +7; h) +3; i) +5; j) +7; k) +5;
             l) +5; m) +5; n) +6; o) +3; p) +4; q) +5; r) +3; s) -3; t) +2; u) 0; v) -2;
             w) +4; x) 0; y) +2; z) 0

Problem II:
a. Na is oxidized and C is reduced.
   Na is the reducing agent and CO_2 is the oxidizing agent.
b. C is being oxidized and O is being reduced.
   C_2H_4 is the reducing agent and O_2 is the oxidizing agent.
c. Cu is being oxidized and Ag is being reduced.
   Cu is the reducing agent and Ag^+ is the oxidizing agent.
d. C is being oxidized and I is being reduced.
   CO is the reducing agent and I_2O_5 is the oxidizing agent.
e. S is being oxidized and Cr is being reduced.
   SO_2 is the reducing agent and CrO_3 is the oxidizing agent.

Self-Test:

1. c
2. a
3. c
4. c
5. c
6. b, c
7. d
8. a
9. a
10. a
11. b
12. a
13. a
14. +7
15. +4
16. +3
17. a) oxidizing agent
    b) oxidized
18. a) oxidized
    b) oxidizing agent
19. c
20. a) -2
    b) -1
    c) oxidized

Chapter 9 Solutions

To understand the subject of solutions, one must learn the language of solutions. Collected in problems 1 and 13 at the end of chapter 9 are most of the new terms introduced in this chapter. Before proceeding you should become familiar with these terms.

We propose to compare the use of some of the terms here. For example, consider soluble/insoluble and miscible/immiscible. These sets of terms are similar in meaning, but miscible indicates that the range of solubility is much greater. To be precise, miscible says solute and solvent can be mixed in all proportions, whereas soluble simply indicates that a significant amount of solute dissolves in the solvent. Miscible/immiscible are terms usually used in describing liquid/liquid systems, although this limitation is by custom and not by definition.

If one wishes to be somewhat more precise as to the amount of solute dissolved in a particular solution, the terms unsaturated, saturated and supersaturated can be employed. These terms would never be used for a set of miscible liquids since they presume that a limited amount of solute can be dissolved in the solvent at a given temperature. Except for saturated, these terms, too, are quite imprecise, indicating simply that the solvent is holding less than it can or more than it should (at equilibrium). The description of a solution as unsaturated can be modified by the use of the terms dilute and concentrated, these latter terms giving a rough idea of how close one is to a saturated solution. (Note that there are some chemists who would describe even a saturated solution as dilute if the solute was one of very low solubility. Again, this is a matter of individual habit.)

If one wishes to go beyond these imprecise descriptions of solutions, then quantitative measurements are called for. Of the terms mentioned above, saturated is the only truly quantitative one. For a particular solute/solvent system at a given temperature, saturated specifies a precise ratio of solute to solvent. Numerically, the concentration of a saturated solution is usually given in grams of solute per 100 g of solvent. Note that the denominator here is grams of solvent. In per cent concentrations, molarity, etc., the denominator refers to amount of solution not solvent. It is permissible to state the concentration of a saturated solution in molarity or weight per cent, etc. We are just noting that most chemical handbooks use grams of solute/100 g solvent.

To complete our review of terms dealing with concentrations, let's take note of two differences between per cent concentrations and molarity. First, since the molarity involves moles and moles involve formula weights, the formula of the solute being considered will make a difference in calculations of molarity. Per cent concentrations ignore what the solute is. A second difference between molarity and per cent concentrations lies in the size of the solution sample being referred to. A 1 molar solution contains 1 mole per ℓ (or 1000 ml). A 1 per cent solution contains 1 g per 100 g or 1 ml per 100 ml. Let's relate the two forms for reporting concentrations with an example. (The formula weight of NaOH = 40.)

$$1 \underline{M} \text{ NaOH} = \frac{1 \text{ mole NaOH}}{1 \ell \text{ of soln}} = \frac{40 \text{ g NaOH}}{1 \ell \text{ of soln}} = \frac{40 \text{ g NaOH}}{1000 \text{ ml of soln}} = \frac{4 \text{ g NaOH}}{100 \text{ ml of soln}} \approx \frac{4 \text{ g NaOH}}{100 \text{ g of soln}}$$

↑ Concentration in molarity

↑ Concentration in weight per cent

The quantitative treatment of solutions is best learned through problems.

Remember:

$$\underline{M} = \text{molarity} = \frac{\text{moles of solute}}{\text{litre of solution}} \qquad \text{moles of solute} = \frac{\text{g of solute}}{\text{gram formula weight of solute}}$$

$$\text{weight \%} = \frac{\text{g of solute}}{\text{g of solution}} \times 100 = \frac{\text{g of solute}}{\text{g of solute} + \text{g of solvent}} \times 100$$

$$\text{volume \%} = \frac{\text{volume of solute}}{\text{volume of solution}} \times 100 = \frac{\text{volume of solute}}{\text{volume of solute} + \text{volume of solvent}} \times 100$$

Problems

I. Calculate the molarity of the solution which results if the stated amount of solute is dissolved in 1 ℓ of solution.
   a. 58.5 g NaCl
   b. 0.4 g NaOH
   c. 4.9 g H_2SO_4
   d. 40 g of $CaBr_2$
   e. 74.5 g $(NH_4)_3PO_4$

II. Calculate the molarity of the following.
   a. 58.5 g NaCl in 2 ℓ of solution
   b. 0.4 g of NaOH in 50 ml of solution
   c. 4.9 g of H_2SO_4 in 50 ml of soln
   d. 40 g of $CaBr_2$ in 0.4 ℓ of solution
   e. 74.5 g of $(NH_4)_3PO_4$ in 10 ℓ of solution

III. How many moles of solute are present in each of the following?
   a. 1 ℓ of 2 $\underline{M}$ NaCl
   b. 2 ℓ of 5 $\underline{M}$ NaOH
   c. 2 ml of 0.1 $\underline{M}$ H_2SO_4
   d. 5 ml of 0.64 $\underline{M}$ $CaBr_2$
   e. 10 ℓ of 0.01 $\underline{M}$ $(NH_4)_3PO_4$

IV. How many grams of solute are present in 1 ℓ of each of the following solutions?
   a. 2 $\underline{M}$ NaCl
   b. 5 $\underline{M}$ NaOH
   c. 0.1 $\underline{M}$ H_2SO_4
   d. 0.64 $\underline{M}$ $CaBr_2$
   e. 0.01 $\underline{M}$ $(NH_4)_3PO_4$

V. How many grams of solute are there in each of the following?
   a. 50 ml of 2 $\underline{M}$ NaCl
   b. 2 ℓ of 5 $\underline{M}$ NaOH
   c. 2 ml of 0.1 $\underline{M}$ H_2SO_4
   d. 5 ml of 0.64 $\underline{M}$ $CaBr_2$
   e. 10 ℓ of 0.01 $\underline{M}$ $(NH_4)_3PO_4$

VI. How many grams of solute are present in 100 g of each of the following solutions?
   a. 10% NaCl
   b. 40% NaOH
   c. 0.5% H_2SO_4
   d. 2% $CaBr_2$
   e. 12% $(NH_4)_3PO_4$

VII. How many grams of solute are present in each of the following?
   a. 40 g of 10% NaCl solution
   b. 1 kg of 40% NaOH solution
   c. 10 ml of 0.5% H_2SO_4 solution
   d. 1 ℓ of 2% $CaBr_2$ solution
   e. 10 g of 12% $(NH_4)_3PO_4$ solution

VIII. What are the molarity, the approximate weight per cent, and the mg % concentrations of a solution in which 0.4 g of NaOH is dissolved in 100 ml of solution?

Problems 3, 9, 10 and 11 at the end of the chapter supplement the above problems.

Osmosis, one of the topics covered in a more qualitative way in this chapter, occasionally presents some difficulties to students who are new to the concept. Usually, one has a pretty clear picture of the process of osmosis itself. Water molecules move back and forth through a semipermeable membrane. If the concentration of water is higher on one side of the membrane than on the other, more water molecules move from the high water-concentration side to the low water-concentration side. In osmosis (as well as in dialysis) it is always true that the tendency is to equalize the concentration on the two sides of the membrane.

So what causes all the problems? It's probably the fact that water-concentration is not usually what is reported for a solution. To understand that statement, first consider again the direction of the net flow of water through the semipermeable membrane. The net flow of water is from the solution where water-concentration is higher to the solution where water-concentration is lower. O.K.? O.K.--then in which direction is the net flow of water if the solution on one side of the membrane is 0.9% NaCl and the solution on the other side is 1.5% NaCl? If you just said from the 1.5% solution to the 0.9% solution, you're wrong! That is certainly from higher concentration to lower concentration, but it's not from higher water-concentration to lower water-concentration. When we say the solution is 1.5% NaCl, we also mean it is 98.5% water. A solution which is 0.9% NaCl is also 99.1% water. Now--in which direction would you say the net flow of water occurs? Right--from the 99.1% water solution (0.9% NaCl) to the 98.5% water solution (1.5% NaCl). Thus, in osmosis one of the stumbling blocks to complete understanding is that we normally are given solute concentration and not solvent concentration. That is why, in the chapter, we state that net water flow is from low (solute) concentration to high (solute) concentration. It is just as true to say that net water flow is from high solvent concentration to low solvent concentration.

The reason we've spent all this time reviewing the process of osmosis is that we don't want you to think that dialysis is fundamentally different from osmosis. Answer this question: If, in dialysis, sodium ions and chloride ions can move through the dialyzing membrane, in which direction will they move if the two solutions separated by the membrane are 0.9% NaCl and 1.5% NaCl? Just like water molecules, these ions will move from where they are more concentrated to where they are less concentrated. In the case of sodium and chloride ions, however, the stated concentrations for the solutions do refer directly to the concentration of these ions. Therefore, the net flow of the ions is from the 1.5% NaCl solution (high concentration) to the 0.9% NaCl solution (low concentration). The net flow of water is in the opposite direction, i.e., from its higher concentration to its lower concentration.

Before we state a simple rule to keep everything straight in our minds, let's look at one other problem. The term pressure usually suggests "pushing". High pressure means something pushes more strongly. Therefore, when we say a solution has a high osmotic pressure and try to connect that information in our minds with the net direction of the flow of water molecules, we might end up with the following reasoning: the higher the osmotic pressure, the more water molecules are pushed out. Wrong! Osmotic pressure was defined in such a way that it measures the tendency of water molecules to move into the solution, not out of the solution. The higher the osmotic pressure of a solution, the greater the tendency for water to be drawn into the solution. The net flow of water is from solutions of low osmotic pressure to solutions of high osmotic pressure.

Now, here's the rule to keep everything straight. (You'll immediately notice it's not a rule but the briefest possible summary of the above discussion.)

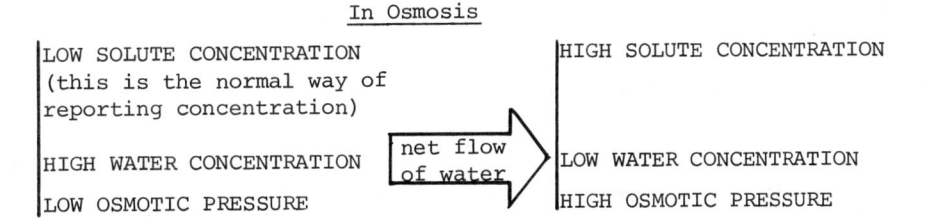

In Osmosis

LOW SOLUTE CONCENTRATION (this is the normal way of reporting concentration)		HIGH SOLUTE CONCENTRATION
HIGH WATER CONCENTRATION	net flow of water ⟹	LOW WATER CONCENTRATION
LOW OSMOTIC PRESSURE		HIGH OSMOTIC PRESSURE

In Dialysis

HIGH SOLUTE (Ion) CONCENTRATION   net flow of ions ⟹   LOW SOLUTE CONCENTRATION

You can check your understanding of the rest of the material covered in this chapter by going through the self-test.

Self-Test - Select the best answer.

1. If a small amount of sugar is dissolved in a large amount of water, sugar is the
   a. solute   b. solvent   c. solution
2. A mixture is homogeneous if
   a. the components of the mixture are so intimately mixed that all samples of the mixture have the same composition
   b. the mixture has a lower osmotic pressure than physiological saline
   c. the mixture will absorb water from the air
3. If the energy required to break the crystal lattice is greater than the energy of solvation,
   a. the solid dissolves in the solvent
   b. the solid is insoluble in the solvent

4. Generally, if a seed crystal is added to a supersaturated solution
   a. the crystal dissolves
   b. the crystal causes all of the solute to precipitate from solution
   c. solute precipitates until an unsaturated solution forms
   d. some solute precipitates and equilibrium is established with the saturated solution
5. An increase in temperature usually results in __?__ of a solid solute.
   a. an increase in the solubility
   b. a decrease in the solubility
   c. an increase in density
   d. an increase in the boiling point
6. Oil and water are
   a. miscible   b. immiscible
7. Salts incorporating all but one of the following ions are usually soluble. Which ion is the exception?
   a. NO_3^- b. SO_4^{2-} c. Na^+ d. NH_4^+
8. Which term is used to describe a compound which loses water on standing in dry air?
   a. efflorescent   b. deliquescent   c. hygroscopic
9. Which is least likely to dissolve in water?
   a. ionic compound   b. polar compound   c. nonpolar compound
10. The compound $Na_2CO_3 \cdot 7H_2O$ is a(an)
    a. hydroxide   b. hydrate   c. anhydrous compound
11. Which is generally not true? An increase in temperature increases
    a. the rate of a reaction
    b. the solubility of a salt in water
    c. the solubility of a gas in water
12. Which compound would not be expected to dissolve in water?
    a. $CH_2CH_2CH_3$ / OH
    b. $CH_3CH_2CH_2$ / NH_2
    c. $CH_3CH_2CH_3$
    d. $CH_2CH_2CH-CH_2CH_2$ / OH OH OH
13. The solubility of sodium chloride in water is 36^{20}. At the same temperature, therefore, a solution of 20 g NaCl in 1 liter of water would be classified as
    a. dilute
    b. concentrated
    c. saturated
    d. supersaturated
14. If a 36% solution of NaCl is saturated, then at the same temperature 10 ml of a solution containing 3 g of NaCl is
    a. dilute   b. concentrated   c. saturated   d. supersaturated
15. If a solution is saturated when its concentration is 5%, then at the same temperature 200 g of the solution would be supersaturated if it contained __?__ grams of solute.
    a. 2.5   b. 5   c. 10   d. 12
16. If the concentration of sodium hydroxide is 10%, then the amount of NaOH in 500 ml of solution is
    2 g   4 g   5 g   10 g   20 g   25 g   40 g   50 g   100 g   500 g
17. In 500 ml of a 0.25 M solution of NaOH are how many grams of NaOH?
    2   4   5   10   20   25   40   50   100   500
18. How many millimoles of HCl are there in 1 ℓ of 0.01 M HCl solution?
    10   10^{-1} 10^{-2} 10^{-3} 10^{-4} 10^{-5}
19. What is the formula weight of a compound if 10 g of the compound in 500 ml of solution gives a 0.5 M solution?
    10   20   40   50   60   80   100
20. A solution containing 100 mg of solute in 1 ℓ of solution has a concentration of
    a. 1 mg%   b. 10 mg%   c. 100 mg%   d. 1000 mg%
21. Which equality is correct?
    a. 1 ppm = 1000 ppb   b. 1000 ppm = 1 ppb
22. A solution which has been diluted 1:3 is less concentrated than the solution which has been diluted
    a. 1:2   b. 1:30   c. 1:3000

23. Compared to pure water, a solution containing one mole of sugar per kg of water
    a. melts at and boils at a higher temperature
    b. melts at and boils at a lower temperature
    c. melts higher and boils lower
    d. melts lower and boils higher
24. The freezing point in °C of a solution which contains 2 moles of sugar per kg of water is
    a. -2   b. 2   c. 80   d. -80   e. -3.72   f. 1.02   g. -1.02   h. 103.72
25. Which solution has the higher osmotic pressure?
    a. 0.1 M NaCl   b. 0.5 M NaCl
26. The chief distinction between dialysis and osmosis is
    a. the size of the particles which pass through the membrane
    b. the direction in which water moves through the membrane
    c. the speed with which the particles pass through the membrane
27. The net flow of water through a semipermeable membrane is
    a. from a solution of higher osmotic pressure to one of lower osmotic pressure
    b. from a solution of lower osmotic pressure to one of higher osmotic pressure
28. A solution contains particles of three sizes (., •, ●). If two of these types of particles pass through a dialyzing membrane, the two must be
    a. . and •   b. . and ●   c. • and ●
29. In both dialysis and osmosis which particles do not pass through the membrane?
    a. water   b. small molecules   c. colloids
30. If two solutions with concentrations of 0.1 M sugar and 0.5 M sugar respectively are separated by a semipermeable membrane, during osmosis there is a net flow of
    a. sugar molecules from the 0.1 M solution to the 0.5 M solution
    b. sugar molecules from the 0.5 M solution to the 0.1 M solution
    c. water molecules from the 0.1 M solution to the 0.5 M solution
    d. water molecules from the 0.5 M solution to the 0.1 M solution
31. Red blood cells undergo hemolysis if placed in a solution which has a __?__ osmotic pressure than the solution inside the blood cells.
    a. higher   b. lower
32. Crenation of cells occurs when the cells are placed in a(an) __?__ solution.
    a. hypotonic   b. isotonic   c. hypertonic   d. physiological saline
33. Which is not true of a colloidal dispersion?
    a. Particles may have a diameter of 10 nm.
    b. The dispersion exhibits the Tyndall effect.
    c. The dispersion does not settle out on standing.
    d. The dispersed particles can be filtered by slowly passing the dispersion through filter paper.
    e. All of the above are true of colloidal dispersion.
34. Which type of mixture cannot exist as a colloidal dispersion?
    a. solid in solid   b. liquid in liquid   c. gas in gas
35. An emulsifying agent is used to
    a. bring about precipitation of solute from a true solution
    b. stabilize a colloidal dispersion
    c. filter the particles of a suspension
36. In the Tyndall effect, one observes
    a. the precipitation of solute triggered by addition of a seed crystal
    b. particles slowly settling out of a suspension
    c. a beam of scattered light
37. Which of the following mixtures will separate on standing?
    a. true solution   b. suspension   c. colloidal dispersion
38. If the BOD of a body of water is high
    a. conditions are excellent for fish life
    b. the water is relatively free of degradable organic matter
    c. microorganisms in the water require large amounts of oxygen to oxidize sewage

ANSWERS

Problems:
  I.   a. 1 M;   b. 0.01 M;   c. 0.05 M;   d. 0.2 M;   e. 0.5 M
  II. a. 0.5 M;   b. 0.2 M;   c. 1 M;   d. 0.5 M;   e. 0.05 M
  III. a. 2 moles;   b. 10 moles;   c. 0.0002 mole;   d. 0.0032 mole;   e. 0.1 mole

IV. a. 117 g; b. 200 g; c. 9.8 g; d. 128 g; e. 1.49 g
V. a. 5.85 g; b. 400 g; c. 0.0196 g; d. 0.64 g; e. 14.9 g
VI. a. 10 g; b. 40 g; c. 0.5 g; d. 2 g; e. 12 g
VII. a. 4 g; b. 400 g; c. 0.05 g approximately; d. 20 g approximately; e. 1.2 g
VIII. 0.1 $\underline{M}$, 0.4%, 400 mg%

Self-Test:

1. a
2. a
3. b
4. d
5. a
6. a
7. b
8. a
9. c
10. b
11. c
12. c
13. a
14. b
15. d
16. 50 g
17. 5
18. 10
19. 40
20. b
21. a
22. a
23. d
24. e
25. b
26. a
27. b
28. a
29. c
30. c
31. b
32. c
33. d
34. c
35. b
36. c
37. b
38. c

Chapter 10 Acids and Bases

Before we do anything else, let's quickly review the definitions of acids. Acids were defined in the chapter in three ways: as compounds which yield hydrogen ions, as compounds which yield hydronium ions in aqueous solutions, and as compounds which act as proton donors. It is now generally accepted that a hydrogen ion does not exist in solution as an independent unit. Thus the second and third definitions are attempts to be a bit more accurate in describing the action of an acid.

A hydrogen ion and a proton are identical species. The Bohr picture of a hydrogen atom is:

$$1p \qquad 1e$$

A hydrogen ion is the hydrogen atom minus its outer shell of electrons. For hydrogen, with only one electron, that means the ion is simply the hydrogen nucleus. The hydrogen nucleus contains a single proton. Therefore, $H^+ = p$. The preferred term for this species is proton, but the symbol most commonly used is H^+.

As we said, the proton or hydrogen ion does not exist as an independent species in solution. In aqueous solutions, protons from acids are transported by water molecules. This is where the hydronium ion comes in. It's like transferring food from your plate to your mouth. It's the food that's transferred, but a fork carries it from one place to another. In reactions involving acids in aqueous solutions, it's the proton being transferred, but a water molecule carries the proton from one place to another.

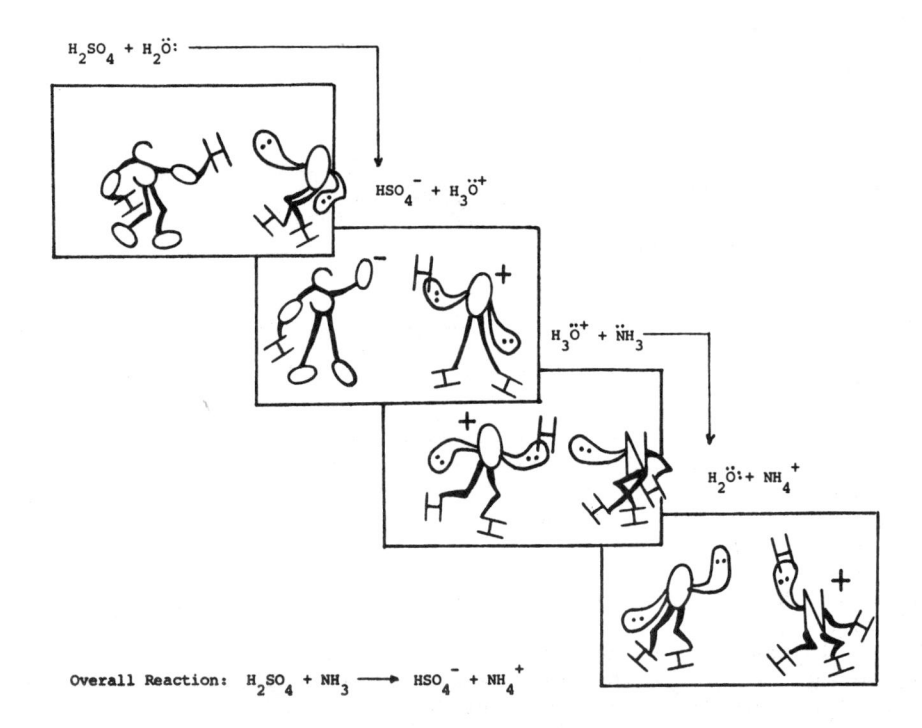

$H_2SO_4 + H_2\ddot{O}: \longrightarrow HSO_4^- + H_3\ddot{O}^+$

$H_3\ddot{O}^+ + \ddot{N}H_3 \longrightarrow H_2\ddot{O}: + NH_4^+$

Overall Reaction: $H_2SO_4 + NH_3 \longrightarrow HSO_4^- + NH_4^+$

The many reactions discussed in this chapter can be classified under just a few headings. First, there are what we could call "the defining equations." These are the equations which say "this compound is an acid" or "this compound is a base." Since acids and bases were defined in a number of ways in the chapter, the defining equation can be written in a number of ways.

Equation for an Arrhenius acid: $HCl \rightleftharpoons H^+ + Cl^-$

Equation for a Brønsted-Lowry acid: $HCl + H_2O \rightleftharpoons H_3O^+ + Cl^-$

Both equations indicate HCl is an acid, but the second one emphasizes that the proton is transferred and not simply released. The extent to which this reaction proceeds to the right determines whether the acid is called strong or weak. Table 10.1 in the text lists several important strong and weak acids and is worth committing to memory.

Equations for an Arrehnius base: $NaOH \longrightarrow Na^+ + OH^-$

$NH_3 + H_2O \rightleftharpoons NH_4^+ + OH^-$

Equations for a Brønsted-Lowry base: $OH^- + H_3O^+ \longrightarrow H_2O + H_2O$

$NH_3 + H_2O \rightleftharpoons NH_4^+ + OH^-$

Notice that the equation for ammonia qualifies under both definitions: it shows the release of hydroxide ion (Arrhenius) and it also shows ammonia picking up a proton (Brønsted-Lowry). For sodium hydroxide, we use two different equations. To satisfy the Arrhenius definition we simply show sodium hydroxide dissociating to produce the hydroxide ion in solution. To satisfy the Brønsted-Lowry definition, we concentrate on the hydroxide ion from sodium hydroxide and show it picking up a proton.

The remaining reactions can be collected into three groups.

Neutralization: Acid plus base yields salt plus water.
$HCl + NaOH \longrightarrow NaCl + H_2O$

Reaction of metal with acid: active metal plus acid yields salt plus hydrogen gas.
$Mg + H_2SO_4 \longrightarrow MgSO_4 + H_2$

Reaction of carbonate (or bicarbonate) with acid:
Acid plus (bi)carbonate yields salt plus carbon dioxide plus water.
$2 HCl + Na_2CO_3 \longrightarrow 2 NaCl + CO_2 + H_2O$
$HCl + NaHCO_3 \longrightarrow NaCl + CO_2 + H_2O$

We note again that the very important reaction of carbonic acid:
$H_2CO_3 \longrightarrow CO_2 + H_2O$

is <u>not</u> a general reaction of acids nor even a general reaction of weak acids. It is a chemical property peculiar to carbonic acid. The equation indicates the special instability of carbonic acid. If this compound is produced in a reaction, e.g.:

$HCl + NaHCO_3 \longrightarrow NaCl + H_2CO_3$

it immediately decomposes:
$\longrightarrow H_2O + CO_2$

Notice that when another weak acid, like acetic acid, is produced in a reaction:
$HCl + CH_3COONa \longrightarrow NaCl + CH_3COOH$

the acid does not decompose. We call your attention to this property of carbonic acid because the equilibra between carbon dioxide, carbonic acid and bicarbonate ion play an extremely important role in controlling the acidity of the blood. This subject will be discussed in chapter 11 and again in chapter 28.

<u>Self-Test</u>

1. If litmus paper changes color from red to blue when placed in an aqueous solution, the solution is:
   a. acidic   b. basic   c. neutral
2. According to the theory of Arrhenius, the properties of acids are the properties of
   a. H^+ b. H_3O^+ c. OH^- d. H_2O e. NH_3
3. According to the theory of Arrhenius, the properties of bases are the properties of
   a. H^+ b. H_3O^+ c. OH^- d. H_2O e. NH_3

4. According to the Brønsted-Lowry theory, an acid is
   a. a proton donor
   b. a proton acceptor
   c. an hydronium ion donor
   d. an hydronium ion acceptor
5. Which is a monoprotic acid?
   a. H_2CO_3 b. H_2SO_4 c. CH_3COOH d. H_3PO_4
6. Which is a weak acid?
   a. HCl   b. HNO_3 c. H_2SO_4 d. H_3PO_4 e. H_3BO_3
7. Which is a strong acid?
   a. H_2CO_3 b. CH_3COOH c. H_3PO_4 d. HNO_3 e. HCN
8. According to the equation, $C_6H_5OH + H_2O \rightleftharpoons C_6H_5O^- + H_3O^+$, the compound C_6H_5OH is a
   a. strong acid   b. strong base   c. weak acid   d. weak base
9. According to the equation, $C_6H_5OH + OH^- \rightleftharpoons C_6H_5O^- + H_2O$, the compound C_6H_5OH is the
   a. stronger acid   b. stronger base   c. weaker acid   d. weaker base
10. The conjugate partner of a strong acid is a
    a. strong acid   b. weak acid   c. strong base   d. weak base
11. According to the equation shown in question 8, which set of compounds constitutes a conjugate acid-base pair?
    a. C_6H_5OH and H_2O b. C_6H_5OH and $C_6H_5O^-$ c. C_6H_5OH and H_3O^+
12. According to the Brønsted-Lowry theory, a weak acid
    a. holds on to its protons relatively tightly
    b. has a very weak hold on its protons
13. The formula for phosphoric acid is
    a. H_2PO_3 b. H_3PO_3 c. H_2PO_4 d. H_3PO_4
14. The name of the acid HBr is
    a. hydrobromic acid
    b. bromic acid
    c. bromous acid
    d. bromate acid
15. If ClO_3^- is the chlorate ion, then $HClO_3$ is
    a. hydrochloric acid
    b. chloric acid
    c. chlorous acid
    d. chlorate acid
16. The metaborate ion is BO_2^-. Orthoboric acid is
    a. BO_2^{3-} b. BO_3^{3-} c. HBO_2 d. H_3BO_3
17. Which is regarded as a strong base in aqueous solution?
    a. KOH   b. $Mg(OH)_2$ c. NH_3
18. Which base exists primarily in un-ionized form in aqueous solution?
    a. KOH   b. $Mg(OH)_2$ c. NH_3
19. Which is not highly ionized in aqueous solution?
    a. NaOH   b. $Ca(OH)_2$ c. H_2SO_4 d. CH_3COOH e. $(NH_4)_3PO_4$
20. This is a portion of the activity series of metals: K, Mg, Cr, H, Hg
    a. Which metal would fail to yield hydrogen gas when treated with an acid?
       K   Mg   Cr   Hg
    b. Which is the strongest reducing agent?
       K   Mg   Cr   H   Hg
    c. Which reaction is more favored according to the activity series?
       1) $Mg^{2+} + Hg \longrightarrow Hg^{2+} + Mg$
       2) $Hg^{2+} + Mg \longrightarrow Hg + Mg^{2+}$

21. Which set of reactants does not produce a gas when mixed?
    a. NaHCO$_3$ and HCl
    b. NaOH and HCl
    c. Mg and HCl
    d. All of the reactions yield a gaseous product.
22. Which is not a product of the reaction: Na$_2$CO$_3$ + HNO$_3$ $\longrightarrow$?
    a. H$_2$ b. Na$^+$ c. CO$_2$ d. H$_2$O   e. NO$_3^-$
23. The equation, H$_2$CO$_3$ $\longrightarrow$ CO$_2$ + H$_2$O, indicates that carbonic acid is
    a. a weak acid
    b. a strong acid
    c. an unstable acid
24. Lye is a common name for
    a. NaOH   b. NH$_3$ c. HCl   d. H$_2$CO$_3$
25. Which equation describes a neutralization reaction?
    a. Cu + 2 Ag$^+$ $\longrightarrow$ 2 Ag + Cu^{2+}
    b. 2 HNO$_3$ + Zn $\longrightarrow$ Zn(NO$_3$)$_2$ + H$_2$
    c. CH$_3$COOH + NaOH $\longrightarrow$ CH$_3$COONa + H$_2$O
    d. All of the equations are for neutralization reactions.
26. The net ionic equation which describes a neutralization reaction is
    a. H$^+$ + OH$^-$ $\longrightarrow$ H$_2$O
    b. HCO$_3^-$ + H$^+$ $\longrightarrow$ H$_2$CO$_3$
    c. H$_2$CO$_3$ $\longrightarrow$ CO$_2$ + H$_2$O
    d. M + 2 H$^+$ $\longrightarrow$ M^{2+} + H$_2$
27. Which process is described by the equation:
    H$_2$SO$_4$(aq) + CaCO$_3$(s) $\longrightarrow$ CaSO$_4$(aq) + H$_2$O + CO$_2$(g)
    a. erosion of limestone by acidic pollutants
    b. neutralization of stomach acid by an antacid
    c. the oxidation of an active metal by an acid
28. Which is not found in antacid preparations?
    a. CaCO$_3$ b. Mg(OH)$_2$ c. NaOH   d. NaHCO$_3$
29. Hydrochloric acid is found
    a. in the stomach   b. in toilet bowl cleaners   c. in both a and b
30. Which solution does not cause severe chemical burns on contact with skin?
    a. concentrated sulfuric acid
    b. concentrated hydrochloric acid
    c. concentrated sodium hydroxide solution
    d. concentrated sodium chloride solution
    e. All cause severe chemical burns.

ANSWERS

1. b
2. a
3. c
4. a
5. c
6. e
7. d
8. c
9. a
10. b
11. b
12. a
13. d
14. a
15. b
16. d
17. a
18. c
19. d
20. a. Hg
    b. K
    c. (2)
21. b
22. a
23. c
24. a
25. c
26. a
27. a
28. c
29. c
30. d

Chapter 11 More Acids and Bases

We offer below a comparison of calculations of moles and molarity and equivalents and normality. Sulfuric acid is used as the example, and we propose to calculate the concentration of a solution containing 9.8 g of H_2SO_4 in a total of 2 ℓ of solution.

<u>Moles and Molarity</u>

Formula Weight: 98

Gram Formula Weight (GFW): 98 g/mole

To calculate number of moles: $\dfrac{\text{weight}}{\text{GFW}}$

Example: $\dfrac{9.8 \text{ g}}{98 \text{ g/mole}} = 0.1$ mole

To calculate molarity: $\dfrac{\text{moles of solute}}{\text{litres of solution}}$

Example: $\dfrac{0.1 \text{ mole}}{2 \text{ ℓ}} = 0.05 \underline{M} \; H_2SO_4$

<u>Equivalents and Normality</u>

Equivalent Weight = $\dfrac{\text{Formula Weight}}{2}$: 49

Gram Equivalent Weight (GEW): 49 g/equivalent

To calculate number of equivalents: $\dfrac{\text{weight}}{\text{GEW}}$

Example: $\dfrac{9.8 \text{ g}}{49 \text{ g/equivalent}} = 0.2$ equivalent

To calculate normality: $\dfrac{\text{equivalents of solute}}{\text{litres of solution}}$

Example: $\dfrac{0.2 \text{ equivalent}}{2 \text{ ℓ}} = 0.1 \underline{N} \; H_2SO_4$

We could have calculated the molarity and then used the relationship

$$\underline{N} = 2 \times \underline{M}$$

to determine the normality. The <u>2</u> in that relationship and the <u>2</u> in the denominator of the calculation of equivalent weight above are both derived from the fact that sulfuric acid is a <u>diprotic</u> acid, i.e., it supplies two protons.

The following problems supplement those at the end of the chapter.

Problem I. A. Calculate the formula weight and the equivalent weight of oxalic acid ($H_2C_2O_4$).
B. Calculate the number of equivalents in 9.0 g of oxalic acid; in 450 g; in 1.8 g.
C. Calculate the normality of the following aqueous solutions of oxalic acid: 9.0 g in 1 ℓ of solution; 450 g in 5 ℓ; 1.8 g in 500 ml.
D. What are the molarities of the solutions in part C.

Problem II. In many instances, particularly where the common mineral acids like H_2SO_4, H_3PO_4, HNO_3, etc., are involved, solutions of the acids are prepared by diluting the concentrated acids. Review examples 11.6, 11.7, and 11.8 in the chapter, then try some of the following.
A. Concentrated H_2SO_4 is 36 <u>N</u>. What volume of this acid should be used in preparing 500 ml of 12 <u>N</u> H_2SO_4? 2 ℓ of 3.6 <u>N</u> H_2SO_4? 100 ml of 9 <u>M</u> H_2SO_4?
B. What concentration of solution results when concentrated H_2SO_4 is diluted as indicated? 10 ml diluted to 100 ml; 0.5 ℓ diluted to 2.5 ℓ; 100 ml diluted to 0.2 ℓ

Problem III. Determine the unknown concentration from the given titration data.
A. If 20 ml of 0.5 <u>N</u> HCl was required to titrate 50 ml of NaOH, what is the normality of the NaOH?
B. If 15 ml of 0.4 <u>N</u> H_2SO_4 was required to titrate 45 ml of NaOH, what is the normality of the NaOH?
C. If 30 ml of 0.4 <u>M</u> H_2SO_4 was required to titrate 60 ml of NaOH, what is the molarity of the NaOH?
D. If 100 ml of 0.1 <u>N</u> H_2SO_4 was required to titrate 200 ml of $Ba(OH)_2$, what is the concentration of the barium hydroxide solution in normality? in molarity?

The other major concept involving calculations which was introduced in this chapter is the pH scale. We could play all sorts of games with pH, like calculating the pH of a 0.0385 $\underline{M}$ HCl solution. But what we really want to do is become comfortable with the concept. For that reason, we'll stick to calculations involving simple powers of ten like 0.001 or 10^{-3} so you can clearly see the pattern for conversion from concentration units to pH or pOH units. You should be able to interconvert among pH, $[H^+]$, pOH and $[OH^-]$ by using the definition of pH (or pOH) and the ion product constant of water. Thus, if you know $[H^+]$ equals 10^{-6} $\underline{M}$, then you also know that the pH is 6, the pOH is 8 (because pH + pOH = 14), and $[OH^-]$ equals 10^{-8} $\underline{M}$. Similarly:

If $[OH^-] = 10^{-3}$ $\underline{M}$, then pOH = 3, pH = 11, $[H^+] = 10^{-11}$ $\underline{M}$.

If pH = 9, then $[H^+] = 10^{-9}$ $\underline{M}$, pOH = 5, $[OH^-] = 10^{-5}$ $\underline{M}$.

If pOH = 1, then $[OH^-] = 10^{-1}$ $\underline{M}$, pH = 13, $[H^+] = 10^{-13}$ $\underline{M}$.

For practice, fill in the blanks in the table.

	pH	pOH	$[H^+]$	$[OH^-]$	acidic or basic?
A.	3	___	___	___	___
B.	___	8	___	___	___
C.	___	___	10^{-12} $\underline{M}$	___	___
D.	___	___	___	0.01 $\underline{M}$	___

Although pOH is as easy to calculate as pH, the latter value is almost always quoted. Therefore, you will find it most useful to gear your thinking to the pH scale. Remember that low pH means there are lots of protons available and that at high pH there are few protons available. Low pH means acidic, high pH means basic. The variations in pH, pOH, $[H^+]$, and $[OH^-]$ can be summarized as follows.

<u>Acidic</u>   Neutral   <u>Basic</u>

$1 \longleftarrow \underset{7}{pH} \longrightarrow 13$

$10^{-1} \longleftarrow \underset{10^{-7}}{[H^+]} \longrightarrow 10^{-13}$

$10^{-13} \longleftarrow \underset{10^{-7}}{[OH^-]} \longrightarrow 10^{-1}$

$13 \longleftarrow \underset{7}{pOH} \longrightarrow 1$

<u>Self-Test</u>

1. If 4 equivalents of a substance are dissolved in 500 ml of solution, the concentration is
    a. 1.25 $\underline{N}$ b. 2 $\underline{N}$ c. 4 $\underline{N}$ d. 8 $\underline{N}$ e. 20 $\underline{N}$
2. If 2 g of an acid in 1 ℓ of solution gives a 0.1 $\underline{N}$ solution, what is the equivalent weight of the acid?
    1   2   4   10   20   40   100   200   400
3. If a sample contains 1 milliequivalent of H_2SO_4, how many grams of H_2SO_4 does it contain?
    0.0049   0.0098   0.0196   0.049   0.098   0.196   49   98   196   49000   98000
4. Twenty milliequivalents of a compound per litre of solution gives a solution with a concentration of
    a. 20 $\underline{N}$ b. 2 $\underline{N}$ c. 0.2 $\underline{N}$ d. 0.02 $\underline{N}$ e. 0.002 $\underline{N}$
5. If 24 g of H_3X in 1 ℓ of solution gives a 3 $\underline{N}$ solution, then the equivalent weight of H_3X is
    3   6   8   12   16   24   48   72
6. When used with respect to sulfuric, hydrochloric or nitric acid, the term dilute refers to a concentration of
    a. 1 $\underline{M}$ b. 1 $\underline{N}$ c. 6 $\underline{M}$ d. 6 $\underline{N}$ e. 12 $\underline{M}$ f. 36 $\underline{N}$
7. A standard base is
    a. sodium hydroxide
    b. a solution of base of known concentration
    c. a basic solution with pH 7

8. A solution is considered strongly acidic if its pH is
      2   6   7   8   13
9. What is the pH of a 0.0001 M solution of HCl?
      10^4 10^{-4} 10^{10} 10^{-10} 4   -4   10   -10
10. What is the pH of a 0.001 M solution of NaOH?
       10^3 10^{-3} 10^{11} 10^{-11} 3   -3   11   -11
11. What is the OH^- concentration in a solution with pH 4?
    a. 10 M   b. 4 M   c. 10^{-10} M   d. 10^{-4} M
12. Which is classified as a basic solution? One for which
    a. $[OH^-] = 10^{-8}$ M   b. $[H^+] = 10^{-4}$ M
    c. pH = 7   d. pH = 9
13. Consider these solutions: (1) 0.001 M NaOH
                                  (2) a solution with pH 12
                                  (3) a solution with pOH 6
                                  (4) 10^{-5} M HCl
    If the solutions are arranged left to right in the order of increasing acidity, the order is
    a. 1 2 3 4
    b. 4 3 2 1
    c. 1 4 3 2
    d. 4 3 1 2
    e. 2 1 3 4
14. Which of the following is an indicator?
    a. a mixture of a weak acid and its salt
    b. litmus
    c. a burette
15. In a titration an indicator is used to
    a. buffer the solution
    b. make the end point of the titration visible
    c. neutralize the acid or base present
16. Which of the following salts when dissolved gives a solution with a pH greater than 7?
    a. NH_4Cl b. $(NH_4)_2SO_4$ c. KBr   d. KCN
17. If the pH of a solution of a salt is 6, the salt must be one which was formed from
    a. a strong acid and a strong base
    b. a strong acid and a weak base
    c. a weak acid and a strong base
    d. a weak acid and a weak base
18. A solution of the salt obtained from the titration of acetic acid with potassium hydroxide would have a pH that is
    a. somewhat acidic
    b. somewhat basic
    c. neutral
19. An aqueous solution of NH_4NO_3 is slightly acidic because of the following equilibrium:
    a. $H_2O \rightleftharpoons H^+ + OH^-$
    b. $NO_3^- + H_2O \rightleftharpoons HNO_3 + OH^-$
    c. $NH_4^+ + H_2O \rightleftharpoons NH_3 + H_3O^+$
20. A buffer is
    a. used to establish the end point of a titration
    b. used to maintain the pH of a solution
    c. a combination of a strong acid and a strong base
21. If a solution is buffered at pH 6, then addition of a small amount of base will result in a pH of approximately
       4   6   8
22. Addition of a small amount of acid to a solution buffered with $HPO_4^{2-}/H_2PO_4^-$ results in the following shift in equilibrium:
    a. $H^+ + HPO_4^{2-} \rightarrow H_2PO_4^-$ b. $H_2PO_4^- + H^+ \rightarrow H_3PO_4$ c. $H_2PO_4^- \rightarrow H^+ + HPO_4^{2-}$

23. Which of the buffers does not play a major role in stabilizing the pH of the blood?
    a. CH_3COOH/CH_3COO^- b. H_2CO_3/HCO_3^- c. $H_2PO_4^-/HPO_4^{2-}$
24. The blood buffers are most often involved in stabilizing the pH in the presence of metabolically produced
    a. acids   b. bases   c. salts
25. A condition called acidosis is diagnosed when the blood pH
    a. falls below 7.35
    b. rises above 7.45
    c. is anywhere outside the range 7.35-7.45

ANSWERS

Problem I.   A. formula weight = 90; equivalent weight = 45
             B. 0.2; 10; 0.04
             C. 0.2 N; 2 N; 0.08 N
             D. 0.1 M; 1 M; 0.04 M

Problem II. A. 167 ml; 0.2 ℓ; 50 ml
             B. 3.6 N or 1.8 M; 7.2 N or 3.6 M; 18 N or 9 M

Problem III. A. 0.2 N; B. 0.13 N; C. 0.4 M; D. 0.05 N, 0.025 M

pH Table	pH	pOH	$[H^+]$	$[OH^-]$	acidic or basic
A.	3	11	$10^{-3} M$	$10^{-11} M$	acidic
B.	6	8	$10^{-6} M$	$10^{-8} M$	acidic
C.	12	2	$10^{-12} M$	$10^{-2} M$	basic
D.	12	2	$10^{-12} M$	0.01 M	basic

Self-Test

1. d
2. 20
3. 0.049
4. d
5. 8
6. d
7. b
8. 2
9. 4
10. 11
11. c
12. d
13. e
14. b
15. b
16. d
17. b
18. b
19. c
20. b
21. 6
22. a
23. a
24. a
25. a

Chapter 12 Electrolytes

For anyone new to the subject, electrolytes seem to cause more confusion when they are weak than when they are strong. The confusing culprit is usually the solubility product relationship, a concept introduced with weak electrolytes.

If you found that, while covering the material on solubility products, you were being distracted by the arithmetic involved in multiplying powers of 10, read through Appendix B in the text. The mathematical manipulation of exponential numbers is quite straight forward. If it's causing problems, you probably just need to refresh your memory on the rules for exponential arithmetic.

It is also possible that your first encounter with the formula for solubility product has left you unnecessarily confused. We'll attempt to clarify the concept using examples. The product of the ion concentration for barium sulfate is written:

$$[Ba^{2+}]\ [SO_4^{2-}]$$

The product of the ion concentrations for barium phosphate is:

$$[Ba^{2+}]^3\ [PO_4^{3-}]^2$$

The exponents introduced in the second product are a reflection of the structure of the compound. Each unit of barium sulfate, $BaSO_4$, produces one barium ion (Ba^{2+}) and one sulfate ion (SO_4^{2-}) in solution. Each unit of barium phosphate, $Ba_3(PO_4)_2$, produces three barium ions and two phosphate ions (PO_4^{3-}) in solution. The solubility product is defined in such a way that this difference between the two compounds is taken into account. The simplest way of stating this is to say that the exponents in the solubility product formula are equal to the subscripts which indicate the combining ratio of the ions in the compound formula.

Try writing the solubility product formulas for the following compounds. Note that these problems do not involve arithmetic evaluations. We just want to see which ions are involved and what their exponents are.

   I. A. 1. BaF_2 2. Ag_2S 3. $BaCO_3$ 4. $Ca(OH)_2$ 5. Li_3PO_4
           6. $Mg_3(PO_4)_2$ 7. $MgNH_4PO_4$

      B. 1. copper(I) hydroxide
          2. copper(II) hydroxide
          3. iron(III) phosphate
          4. iron(II) carbonate
          5. calcium phosphate
          6. calcium hydrogen phosphate

Are we interested in the solubility of all of these compounds? No--we just want to make sure that the next time you look at the solubility product for a particular salt you will not wonder where all those exponents are coming from.

For a given salt (at a given temperature, although we are ignoring temperature for the most part) the product of the ion concentrations cannot exceed a certain value, the solubility product constant. Such constants are determined experimentally and are listed in chemical handbooks. The symbol for solubility product constant is K_{sp}. You would not calculate a solubility product constant or K_{sp}. That number has to be supplied to you. What we have been doing in this chapter is calculating the product of ion concentrations for various solutions. We then compare this calculated value to the known K_{sp} for a given salt. If the calculated value exceeds the K_{sp}, we conclude that precipitation of the salt occurs. If the calculated value is equal to or smaller than the K_{sp}, no precipitation occurs, and in the latter case, more of the salt can be dissolved in the solution.

One problem here lies in evaluating when one exponential number exceeds or is less than another. In problems involving K_{sp}, we routinely compare numbers with negative exponents. Remember that the value of a number with a large negative exponent is smaller than the value of a number with a small negative exponent.

$$10^{-12} \text{ is smaller than } 10^{-6}$$
$$1.7 \times 10^{-21} \text{ is larger than } 1.2 \times 10^{-22}$$
$$9.7 \times 10^{-9} \text{ is smaller than } 2.5 \times 10^{-5}$$

And finally, let's take a quick look at what may be one last source of confusion. The solubility product relationship has been defined in such a way that it incorporates exponents for those salts whose formulas indicate a combining ratio involving more than one cation or anion, e.g., $Ca_3(PO_4)_2$. For calcium phosphate, the solubility product is:

$$[Ca^{2+}]^3 [PO_4^{3-}]^2$$

If the concentration of calcium ion in a solution is 0.01 M and the concentration of phosphate ion is adjusted to 0.0001 M, the product of the ion concentrations for calcium phosphate is:

$$(0.01)^3 (0.0001)^2$$

But most scientists would immediately convert those concentrations to exponential form: $0.01 = 10^{-2}$ and $0.0001 = 10^{-4}$. The product of the ion concentrations may then be written:

$$(10^{-2})^3 (10^{-4})^2$$

If you don't use exponential numbers regularly, you may find the appearance of exponents within exponents confusing. Scientists would simply argue that it is easier to raise 10^{-2} to the third power (multiply the exponents to get 10^{-6}). If you raise 0.01 to the third power, you would have to do something like this:

```
                              .01 ⎫
                          x  .01  ⎬ total of 4 decimal places
                            .0001 ⎭→ 4 decimal places
  total of 6 decimal places ⎰  
                            ⎱  x  .01
                6 decimal places ← .000001
```

You get the same answer ($0.000001 = 10^{-6}$), but it probably took longer.

Now--let's take a quick overview of these K_{sp} problems. Consider this specific problem: The K_{sp} for calcium phosphate is 4×10^{-27} (at 37 °C). At the same temperature, will precipitation of calcium phosphate occur from a solution in which calcium ion concentration is 0.01 M and phosphate ion concentration is 0.0001 M?

1) What is the formula for calcium phosphate? $Ca_3(PO_4)_2$
2) What is the solubility product formula for $Ca_3(PO_4)_2$? $[Ca^{2+}]^3 [PO_4^{3-}]^2$
3) What is the product for the given ion concentrations?
   $(0.01)^3 (0.0001)^2$ or $(10^{-2})^3 (10^{-4})^2 = (10^{-6})(10^{-8}) = 10^{-14}$
4) Does this exceed the K_{sp} for calcium phosphate?
   10^{-14} is larger than 4×10^{-27}
   Calcium phosphate will precipitate from this solution.

Sometimes, instead of just stating the ion concentrations as we did above, these concentrations are given indirectly. In these instances, the concentration of the soluble salt which was dissolved to produce the ion in solution is given. We would say 0.01 mole of calcium chloride (which yields 0.01 mole of calcium ion) and 0.0001 mole of sodium phosphate (which yields 0.0001 mole of phosphate ion) were dissolved in 1 ℓ of solution. The results would be the same as given in the example above.

This discussion supplements the material in sections 12.7 and 12.8 in the text. We are not so much interested in your becoming an expert in working solubility product problems as in your understanding the chemical explanation presented in the chapter for bone growth, tooth decay, etc. The following problems are presented for those of you who still want to double check your understanding of the solubility product relationship.

II. A. The K_{sp} of barium carbonate ($BaCO_3$) is 5.1×10^{-9}. Will precipitation of this salt occur from a solution in which both the barium ion and carbonate ion concentrations were originally 0.001 M?

B. The K_{sp} of aluminum sulfide (Al_2S_3) is 2×10^{-7}. Will precipitation occur if the concentration of aluminum ion is 0.1 M and sulfide ion is 0.01 M?

C. A solution is prepared by dissolving 0.1 mole of sodium nitrate ($NaNO_3$) and 0.01 mole of barium chloride ($BaCl_2$) in one litre of the solution. If the K_{sp} of barium nitrate, $Ba(NO_3)_2$, is 4.5×10^{-3}, will this salt precipitate from solution?

Much of the terminology of this chapter (see question 1 at the end of the chapter) was introduced originally in chapter 2. The phenomena of freezing point depression, boiling point elevation and osmotic pressure were originally discussed in chapter 9. Oxidation/reduction (chapter 8) and electron transfer reactions and the activity series of metals (chapter 10) were also considered previously. Thus this chapter ties together many ideas that may have seemed unrelated. Some review of relevant material in the earlier chapters may prove helpful in understanding the material in the present chapter. The questions at the end of this chapter serve as a review for much of its contents. The self-test will further check your understanding of the material.

Self-Test - Select the best answer for each.

1. The soluble salt ammonium acetate is a
    a. strong electrolyte   b. weak electrolyte   c. nonelectrolyte
2. Which is not a strong electrolyte?
    a. NaCl   b. $NaHCO_3$ c. HCl   d. NaOH   e. H_2CO_3
3. Which will cause the electric light of the conductivity apparatus to glow only weakly?
    a. sodium acetate   b. acetic acid   c. sodium hydroxide
4. Which will not conduct an electric current efficiently?
    a. ionic solid   b. metallic solid   c. ionic melt   d. solution of ions
5. Which conducts a current most efficiently in solution?
    a. strong acid   b. weak base   c. insoluble salt   d. sugar
6. If one mole of each of the following salts were dissolved in water, which of the water solutions would show the greatest freezing point depression?
    a. NaCl   b. NH_4NO_3 c. $MgCl_2$ d. K_3PO_4
7. A solution of hydrogen chloride in water conducts electricity because, in water, hydrogen chloride undergoes a process called
    a. dissociation   b. ionization   c. melt formation
8. Soluble salts conduct electricity in solution because they undergo a process called
    a. dissociation   b. ionization   c. melt formation
9. When sodium sulfide dissolves in water it yields two sodium ions and one sulfide ion. Therefore, a solution of sodium sulfide
    a. is not electrically neutral as a whole
    b. is electrically neutral as a whole but still conducts an electric current
    c. is electrically neutral as a whole and does not conduct an electric current
10. The osmotic pressure of a solution is greatest if it contains
    a. one mole of particles whose formula weight is 50
    b. one mole of particles whose formula weight is 400
    c. two moles of particles whose formula weight is 20
    d. All of these solutions would have the same osmotic pressure.
11. Which is not true of nonelectrolytes?
    a. Nonelectrolytes fail to yield ions in solution.
    b. When dissolved, nonelectrolytes have no effect on freezing point, boiling points or osmotic pressure of the solution.
    c. Nonelectrolytes do not conduct an electric current in solution.
12. In the reaction, $MgCl_2 \longrightarrow Mg + Cl_2$, which element is formed at the anode?
    a. Mg   b. Cl_2 c. $MgCl_2$
13. In electroplating, the metal plates out on the
    a. anode
    b. cathode
    c. neither the anode nor the cathode
    d. both the anode and the cathode
14. In electrolysis, oxidation occurs at the
    a. anode   b. cathode
15. Electrical batteries generate a current by taking advantage of
    a. a difference in the tendency of chemical species to give up electrons
    b. the ability of a solution to build up an excess of positive charge
    c. the ability of a solution to build up an excess of negative charge

16. The electrical activity of the brain is measured by
    a. an electrolysis apparatus
    b. a battery
    c. an electrocardiograph
    d. an electroencephalograph
17. A salt will precipitate from solution if the product of the ion concentrations for the salt
    a. exceeds the solubility product constant for the salt
    b. is less than the solubility product constant for the salt
    c. is equal to the solubility product constant for the salt
18. If the solubility product constant for a salt is 3×10^{-9}, precipitation of the salt will occur if the product of the ion concentrations in the solution is
    a. 9.4×10^{-10} b. 2.0×10^{-9} c. 4.6×10^{-8}
19. The concentration of calcium ion in a solution is 10^{-2} M. The concentration of fluoride ion in the solution is 10^{-1} M. What is the ion product for CaF_2 in this solution?
    a. 10^{-1} b. 10^{-2} c. 10^{-3} d. 10^{-4} e. 10^{-5} f. 10^{-6} g. 10^{-7} h. 10^{-8}
20. The solubility product constant for CaF_2 is 2.7×10^{-11}. Will calcium fluoride precipitate from the solution described in question 19?
    a. yes   b. no
21. If bone formation occurs as calcium phosphate precipitates from solution, removal of phosphate ion from the solution will result in
    a. more rapid bone formation
    b. decrease in the rate of bone formation
    c. no change in the precipitation of calcium phosphate
22. The formation of dental caries is favored by
    a. a decrease in the pH at the surface of the tooth
    b. an increase in the pH at the surface of the tooth
    c. an increase in the phosphate ion concentration at the surface of the tooth
23. Sodium chloride is one of the electrolytes which helps to establish the fluid balance in the body by
    a. decreasing the melting point of body fluids
    b. increasing the boiling point of body fluids
    c. by contributing to the osmotic pressure of the body fluids
24. Iodide salts are important to the
    a. oxygen transport system of the body
    b. the functioning of the thyroid gland
    c. proper development of bone and teeth
25. The ions of which element are incorporated in the hemoglobin molecule?
    a. iron b. iodine c. calcium d. phosphorus e. sodium f. chlorine

ANSWERS

<u>Problems I.</u> A. 1. $[Ba^{2+}][F^-]^2$ 2. $[Ag^+]^2[S^{2-}]$ 3. $[Ba^{2+}][CO_3^{2-}]$
4. $[Ca^{2+}][OH^-]^2$ 5. $[Li^+]^3[PO_4^{3-}]$ 6. $[Mg^{2+}]^3[PO_4^{3-}]^2$
7. $[Mg^{2+}][NH_4^+][PO_4^{3-}]$

B. 1. CuOH, $[Cu^+][OH^-]$ 2. $Cu(OH)_2$, $[Cu^{2+}][OH^-]^2$
2. $FePO_4$, $[Fe^{3+}][PO_4^{3-}]$ 4. $FeCO_3$, $[Fe^{2+}][CO_3^{2-}]$
5. $Ca_3(PO_4)_2$, $[Ca^{2+}]^3[PO_4^{3-}]^2$ 6. $CaHPO_4$, $[Ca^{2+}][HPO_4^{2-}]$

<u>Problems II.</u> A. Ion product = $(10^{-3})(10^{-3}) = 10^{-6}$
The ion product is greater than the solubility product constant; the salt will precipitate.

B. Ion product = $(10^{-1})^2(10^{-2})^3 = 10^{-8}$
The ion product is smaller than the solubility product constant; the salt will not precipitate

C. Ion product = $(10^{-2})(10^{-1})^2 = 10^{-4}$
The ion product is smaller than the solubility product constant; the

salt will not precipitate.

<u>Self-Test</u>
1. a
2. e
3. b
4. a
5. a
6. d
7. b
8. a
9. b
10. c
11. b
12. b
13. b
14. a
15. a
16. d
17. a
18. c
19. d
20. a
21. b
22. a
23. c
24. b
25. a

Chapter 13 Bioinorganic Chemistry

There are no major new chemical concepts introduced in chapter 13. We have paused in our theoretical development of the subject to undertake a brief, descriptive survey of several families of elements. We had previously been looking below the surface of matter--developing a picture of its atomic and molecular structure--evaluating phenomena like evaporation or precipitation with models such as the kinetic molecular theory. Now we are looking at various elements as we might actually encounter them in our environment. We are making the point that structural theories were developed, for the most part, to account for observed properties of compounds. For example, the observed chemical inertness of the noble gases led to the development of the octet rule. The unifying theme of this chapter is that chemistry is not merely a classroom subject, but a collection of observable phenomena. Your understanding of the theory of chemistry enhances your appreciation of what you observe around you.

Several groups of compounds were identified by family names in this chapter. The periodic table shown below identifies the groups by these common names.

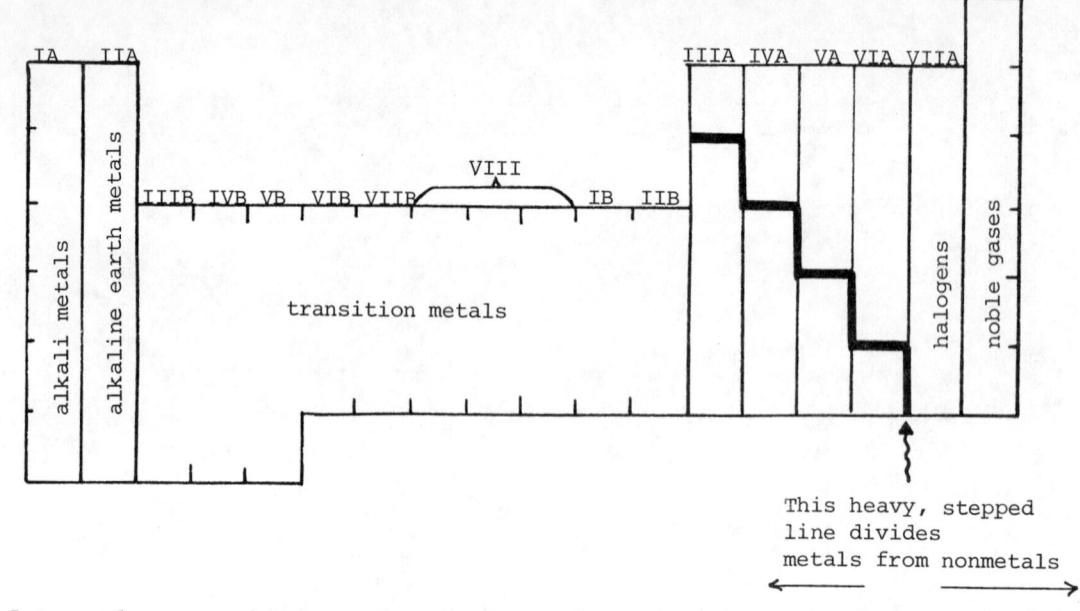

This heavy, stepped line divides metals from nonmetals

Let us also summarize here, for easy comparison, the distinction between conditions encountered in London smog and in photochemical (Los Angeles) smog. (Most real smog episodes involve aspects of both types.)

	London smog	Photochemical smog
Climatic conditions	cool, damp weather	sunny, dry weather
Mode of primary pollutant formation	burning of fuels with high sulfur content (typically coal)	operation of automobiles, i.e., combustion of fuels at high temperatures in, for example, high compression engines
Primary pollutants (formed in the initial combustion process)	sulfur oxides, particulates like soot	nitrogen oxides, carbon monoxide, hydrocarbons
Secondary pollutants (formed by reaction of primary pollutants)	sulfuric acid, ammonium sulfate	ozone, PAN, and many complex organic molecules

	London smog	Photochemical smog
Effect of pollutants	aggravate respiratory problems; attack plant life; corrode construction materials	irritate eyes and respiratory tract; attack crops; corrode metals; cause oxygen deprivation by tying up hemoglobin

The questions at the end of the chapter in the text will lead you through a detailed review of the descriptive material in the chapter. After you've covered those questions, try the self-test.

Self-Test - You may refer to a periodic table.

1. Barium is considered a(an)
   a. alkali metal   b. alkaline earth metal   c. transition metal
   d. noble gas   e. halogen
2. Which is a halogen?
   a. C   b. Ca   c. Cs   d. Cl   e. Co
3. Which is a transition element?
   a. C   b. Ca   c. Cs   d. Cl   e. Co
4. Which is an alkali metal?
   a. C   b. Ca   c. Cs   d. Cl   e. Co
5. Which element is characterized by its extreme nonreactivity?
   a. O_2 b. He   c. Na   d. F_2
6. One of the noble gases is formed in the alpha decay of radioactive elements. Which one?
   a. He   b. Ne   c. Ar   d. Kr   e. Xe   f. Rn
7. The compounds of which element are <u>generally</u> excluded from studies of inorganic chemistry?
   a. B   b. C   c. K   d. Ar   e. U
8. The noble gases have been employed for a number of purposes. Which of the following does not apply to one or more of the noble gases?
   a. used to provide the lift for lighter than air ships
   b. used to dilute oxygen in breathing mixtures
   c. used to provide unreactive atmospheres for various purposes
   d. used to absorb sunlight in the upper atmosphere
9. The glow produced in a neon sign is an example of
   a. the production of a line spectra by excited atoms
   b. photosynthesis
   c. photochemical smog
10. The term "inert gas" is no longer used for the noble gases because
    a. helium can cause the bends
    b. some compounds of the noble gases, like XeF_4, have been formed
    c. the noble gas radon-226 is radioactive
    d. neon signs glow
11. Chlorine and fluorine (or their compounds) are added to drinking water in some areas. Which of the following is <u>not</u> an objective of this water treatment?
    a. the destruction of bacteria
    b. strengthening of tooth enamel
    c. reduction of the incidence of goiter
12. Which hydrohalic acid is involved in the digestive process?
    a. HF   b. HCl   c. HBr   d. HI   e. H_2SO_4
13. Which family of elements forms oxides which are classified as acidic?
    a. alkali metals   b. alkaline earth metals   c. nonmetals
14. Which condition is associated with London rather than photochemical smog?
    a. sunny, dry weather
    b. combustion of high sulfur coal
    c. operation of automobiles
15. Which of the compounds is one of the primary pollutants associated with London-type smog?
    a. NO   b. O_3 c. SO_2 d. PAN

16. The greenhouse effect results from an increase in the concentration of one of the following substances in the atmosphere. Which one?
    a. CO_2 b. NO_2 c. O_3 d. SO_2 e. particulates
17. Because of the greenhouse effect, the temperature of the earth is expected to
    a. increase   b. decrease
18. In which location is ozone considered beneficial to life on earth?
    a. in the upper atmosphere
    b. at ground level
    c. Ozone is extremely toxic and is never considered beneficial to life.
19. Which pollutant does not contribute significantly to the formation of acid rain?
    a. SO_3 b. NO_2 c. CO
20. Which is a combination which exhibits a harmful, synergystic effect?
    a. sulfur dioxide and particulate matter as pollutants
    b. oxygen and nitrogen as a breathing mixture
    c. ozone and aerosol propellants in the upper atmosphere
21. Nitrogen fixation is accomplished by
    a. some bacteria   b. lightning   c. automobile engines
    d. the Haber process   e. all of these   f. none of these
22. Nitrates are used as
    a. fertilizers   b. explosives   c. both a and b   d. neither a nor b
23. Which is not an allotropic form of carbon?
    a. carbon black   b. carbon dioxide   c. diamond   d. graphite
    e. all of the above are allotropic forms of carbon
24. Partial but incomplete combustion of carbon yields
    a. CO   b. CO_2 c. CH_4
25. Of the following toxic gases, which is colorless, odorless and tasteless?
    a. NO_2 b. H_2S c. NH_3 d. CO
26. Which pollutant could be responsible for the observed lowering of the earth's average temperature?
    a. O_3 b. CO_2 c. particulates
27. The salts of which alkali metal are used in the treatment of manic-depressive psychoses?
    a. Li   b. Na   c. K   d. Rb   e. Cs   f. Fr
28. Alkali metal ions
    a. contribute to the osmotic pressure of fluids in living systems
    b. maintain electrical potentials in cells
    c. both a and b
29. The presence of certain ions in water causes the water to be classified as hard. Which ion is not included in this group?
    a. Mg^{2+} b. Fe^{2+} c. Fe^{3+} d. K^+ e. Ca^{2+}
30. Which alkaline earth metal ion is found in chlorophyll?
    a. Be   b. Mg   c. Ca   d. Sr   e. Ba   f. Ra
31. In transition elements
    a. the outer electron shell is a perfect octet
    b. inner electron shells are not completely filled
    c. both of the above are true
32. Which transition metal is not recognized as necessary for good health?
    a. Fe   b. Co   c. Mn   d. Hg

ANSWERS

1. b
2. d
3. e
4. c
5. b
6. a
7. b
8. d
9. a
10. b
11. c
12. b
13. c
14. b
15. c
16. a
17. a
18. a
19. c
20. a
21. e
22. c
23. b
24. a
25. d
26. c
27. a
28. c
29. d
30. b
31. b
32. d

Chapter 14 Hydrocarbons

Much space in chapter 14 has been devoted to the subject of nomenclature. In many ways, nomenclature is a game like Monopoly or poker or baseball. There are rules to be learned and, in the beginning, it sometimes seems that you'll never remember all of them. However, by the time you finish working through questions 4, 5, 6 and 7 at the end of the chapter, you should have a pretty good grasp of the rules. We strongly recommend that you do take the time to work these problems. You'll find that an investment of time now will be paid back in future chapters. The basic rules for naming hydrocarbons provide a foundation for naming all other families of organic compounds.

As we've indicated, the problems at the end of the chapter offer a lot of practice in naming and drawing compounds. For those students who want or need more, we offer the following challenges.

Nomenclature and Isomerism
I. Draw and name the nine isomeric alkanes containing a total of seven carbon atoms.
II. A. Draw and name the thirteen isomeric alkenes (not including geometric isomers) containing a total of six carbon atoms.
   B. Indicate which of the compounds in part A exist as cis/trans isomers.
III. Draw and name the seven isomeric alkynes containing a total of six carbon atoms.
IV. Draw and name each of the five compounds containing a total of five saturated carbon atoms and incorporating a ring.
V. Draw and name the eight compounds containing a benzene ring and 3 additional saturated carbon atoms.

The rest of the chapter's material is reviewed in the self-test.

Self-Test - Select the best answer.
1. The one element necessarily present in every organic compound is
    a. hydrogen   b. oxygen   c. carbon   d. nitrogen   e. sulfur
2. Compounds containing only carbon and hydrogen are known as
    a. methane   b. hydrocarbons   c. carbohydrates   d. isomers   e. aromatic
3. A series of carbon compounds in which each member differs by CH_2 from the preceding member of the series is known as a(n)
    a. aromatic series   b. homologous series   c. hydrocarbon series   d. paraffin series
4. Two of the terms describe the same class of compounds. Which two?
    a. alkanes   b. alkenes   c. alkynes   d. paraffins
5. The simplest hydrocarbon is
    a. CH_2 b. C_2H_4 c. C_2H_6 d. CH_3 e. CH_4
6. Compounds comprised of the same number and kinds of atoms but differing in their arrangement are known as
    a. isotopes   b. isomers   c. homologs   d. allotropes
7. In ethylene the two carbons are joined by a(n)
    a. ionic bond   b. single bond   c. double bond   d. triple bond
8. Restricted rotation about double bonds results in
    a. geometric isomerism   b. fused ring compounds   c. aromatic compounds
9. Which compound does not contain a double bond?
    a. acetylene   b. butene   c. cyclohexene   d. propylene
10. In the name cyclohexane, the prefix "cyclo" means that
    a. the carbon atoms are joined in a ring
    b. the compound is explosive
    c. the compound is a derivative of benzene
    d. each carbon is attached to every other carbon atom
    e. the carbons have a valence of three
11. Which is an unsaturated hydrocarbon?
    a. H-C(H)(Cl)-C(H)(Cl)-H
    b. H_3C-C(H)=O
    c. H-C≡C-H
    d. H-C(H)(H)-C(H)(H)-O-H

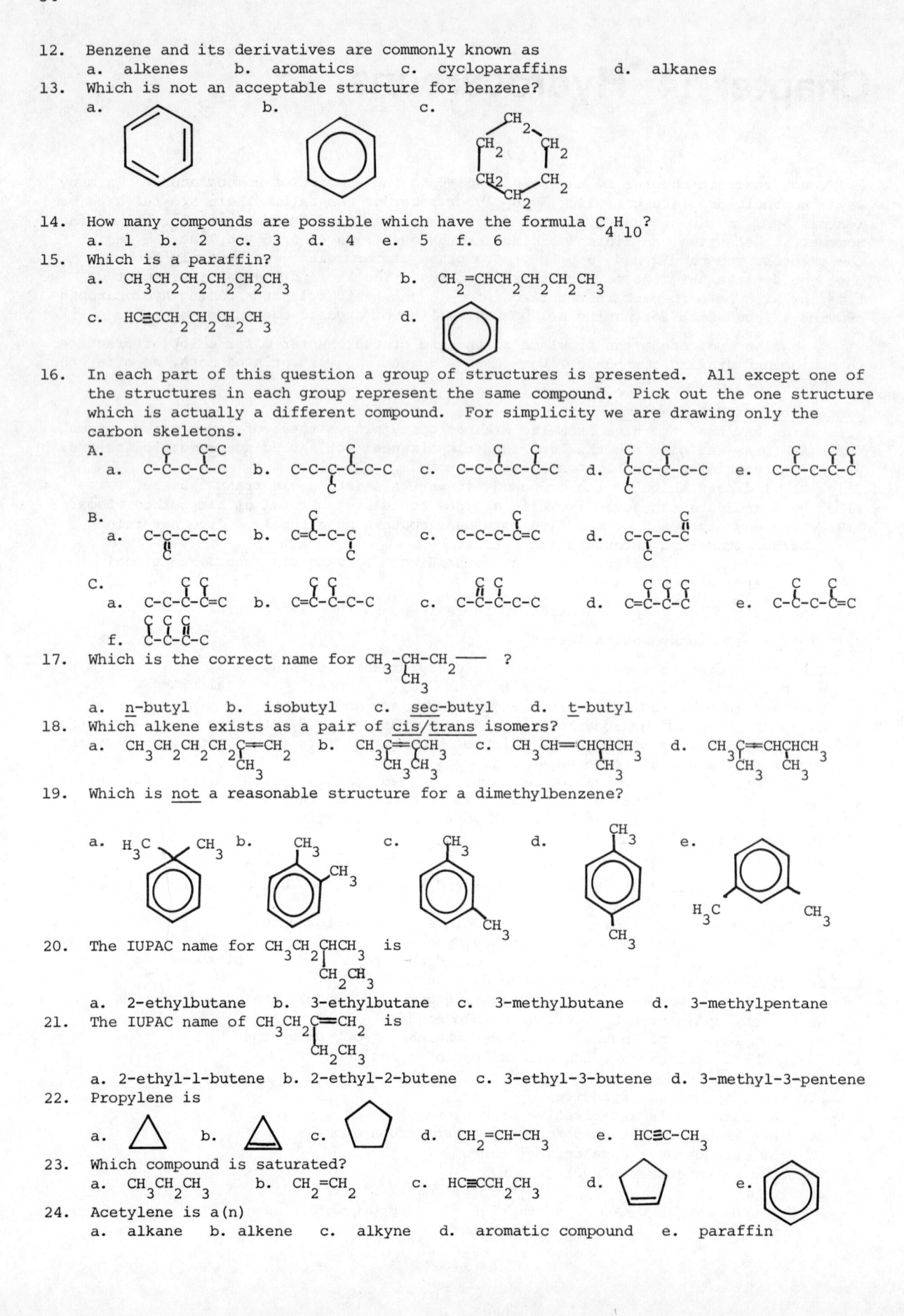

12. Benzene and its derivatives are commonly known as
    a. alkenes   b. aromatics   c. cycloparaffins   d. alkanes
13. Which is not an acceptable structure for benzene?
    a.   b.   c.

14. How many compounds are possible which have the formula C_4H_{10}?
    a. 1   b. 2   c. 3   d. 4   e. 5   f. 6
15. Which is a paraffin?
    a. $CH_3CH_2CH_2CH_2CH_2CH_3$ b. $CH_2=CHCH_2CH_2CH_2CH_3$
    c. $HC\equiv CCH_2CH_2CH_3$ d.

16. In each part of this question a group of structures is presented. All except one of the structures in each group represent the same compound. Pick out the one structure which is actually a different compound. For simplicity we are drawing only the carbon skeletons.

    A.
    a. C-C-C-C-C with C,C-C on top   b. C-C-C-C-C-C with C on top   c. C-C-C-C-C-C with C top and C bottom   d. C-C-C-C-C with C,C top and C bottom   e. C-C-C-C-C with C,C,C on top

    B.
    a. C-C-C-C-C with C,C bottom   b. C=C-C-C with C top, C bottom   c. C-C-C-C=C with C top   d. C-C-C-C with C top, C bottom

    C.
    a. C-C-C-C=C with C,C top   b. C=C-C-C-C with C,C top   c. C-C-C-C-C with C,C,C top   d. C=C-C-C with C,C,C top   e. C-C-C-C=C with C,C top
    f. C-C-C-C with C,C,C top

17. Which is the correct name for $CH_3-CH-CH_2-$ with CH_3 below?
    a. n-butyl   b. isobutyl   c. sec-butyl   d. t-butyl
18. Which alkene exists as a pair of cis/trans isomers?
    a. $CH_3CH_2CH_2CH_2C=CH_2$ with CH_3 b. $CH_3C=CCH_3$ with CH_3,CH_3 c. $CH_3CH=CHCHCH_3$ with CH_3 d. $CH_3C=CHCHCH_3$ with CH_3,CH_3
19. Which is not a reasonable structure for a dimethylbenzene?

    a. H_3C, CH_3 on same carbon of ring   b. 1,2-   c. 1,3-   d. 1,4-   e. 1,3-

20. The IUPAC name for $CH_3CH_2CHCH_3$ with CH_2CH_3 is
    a. 2-ethylbutane   b. 3-ethylbutane   c. 3-methylbutane   d. 3-methylpentane
21. The IUPAC name of $CH_3CH_2C=CH_2$ with CH_2CH_3 is
    a. 2-ethyl-1-butene   b. 2-ethyl-2-butene   c. 3-ethyl-3-butene   d. 3-methyl-3-pentene
22. Propylene is
    a. triangle   b. triangle with double bond   c. pentagon   d. $CH_2=CH-CH_3$ e. $HC\equiv C-CH_3$
23. Which compound is saturated?
    a. $CH_3CH_2CH_3$ b. $CH_2=CH_2$ c. $HC\equiv CCH_2CH_3$ d. cyclopentene   e. benzene
24. Acetylene is a(n)
    a. alkane   b. alkene   c. alkyne   d. aromatic compound   e. paraffin

25. Which is not a gas?
    a. methane   b. ethene   c. acetylene   d. octane
26. If hexane and water are mixed, the result is
    a. a clear solution of hexane dissolved in water
    b. a layer of hexane sitting on top of a layer of water
    c. a layer of water sitting on top of a layer of hexane
27. If two positions on a benzene ring are substituted, how many isomers are possible?
    a. 1   b. 2   c. 3   d. 4   e. 5   f. 6   g. depends on the substituents
28. Two adjacent substituents on a benzene ring are said to be
    a. ortho to one another
    b. meta to one another
    c. para to one another
    d. 1,1-disubstituted
29. Addition reactions are characteristic of
    a. alkanes   b. alkenes   c. aromatic compounds
30. The process in which large hydrocarbon molecules are heated in the absence of air and converted to smaller and more highly branched structures is called
    a. combustion   b. cracking   c. hydration   d. hydrogenation   e. unsaturation
31. Tetraethyl lead is a(n)
    a. gasoline additive   b. anesthetic   c. emollient
32. The raw material from which most hydrocarbons are obtained is
    a. gasoline   b. petroleum   c. animal fats   d. vegetable oils
33. Methane gas can cause death through
    a. chemical pneumonia   b. carcinogenesis   c. asphyxiation
34. Which act as emollients?
    a. gaseous alkanes
    b. low boiling liquid alkanes
    c. high boiling liquid alkanes
35. Which hydrocarbon is used as an effective anesthetic?
    a. methane   b. benzene   c. cyclopropane
36. Hydrogenation of vegetable oils produces
    a. butter   b. saturated fats   c. unsaturated fats
37. A common fused-ring aromatic compound is
    a. benzene   b. naphthalene   c. toluene
38. In each case, select the property which is typical of organic rather than inorganic compounds.
    A. a. melt below 200 °C   b. melt above 200 °C
    B. a. water soluble   b. water insoluble
    C. a. ionic bonding   b. covalent bonding
    D. a. specific gravity less than 1 b. specific gravity greater than 1
    E. a. flammable   b. nonflammable

ANSWERS

Nomenclature and Isomerism - Only the carbon skeleton of structural formulas are given.

I. C-C-C-C-C-C-C   heptane   II. C=C-C-C-C-C   1-hexene
    C-C-C-C-C-C   2-methylhexane   C-C=C-C-C-C   2-hexene (cis and trans)
     |
     C
    C-C-C-C-C-C   3-methylhexane   C-C-C=C-C-C   3-hexene (cis and trans)
        |
        C
         C
         |
    C-C-C-C-C   2,2-dimethylpentane   C=C-C-C-C   2-methyl-1-pentene
      | |
      C   C
      C C
      | |
    C-C-C-C-C   2,3-dimethylpentane   C=C-C-C-C   3-methyl-1-pentene
                                                         |
                                                         C
      C   C
      | |
    C-C-C-C-C   2,4-dimethylpentane   C-C-C-C-C   4-methyl-1-pentene
                                                         |
                                                         C
          C
          |
    C-C-C-C-C   3,3-dimethylpentane   C-C=C-C-C   2-methyl-2-pentene
          | |
          C   C
        C-C
        |
    C-C-C-C-C   3-ethylpentane   C-C=C-C-C   3-methyl-2-pentene (cis and trans)
                                                          |
                                                          C
      C C
      | |
    C-C-C-C   2,2,3-trimethylbutane   C-C=C-C-C   4-methyl-2-pentene (cis and trans)
      | |
      C   C

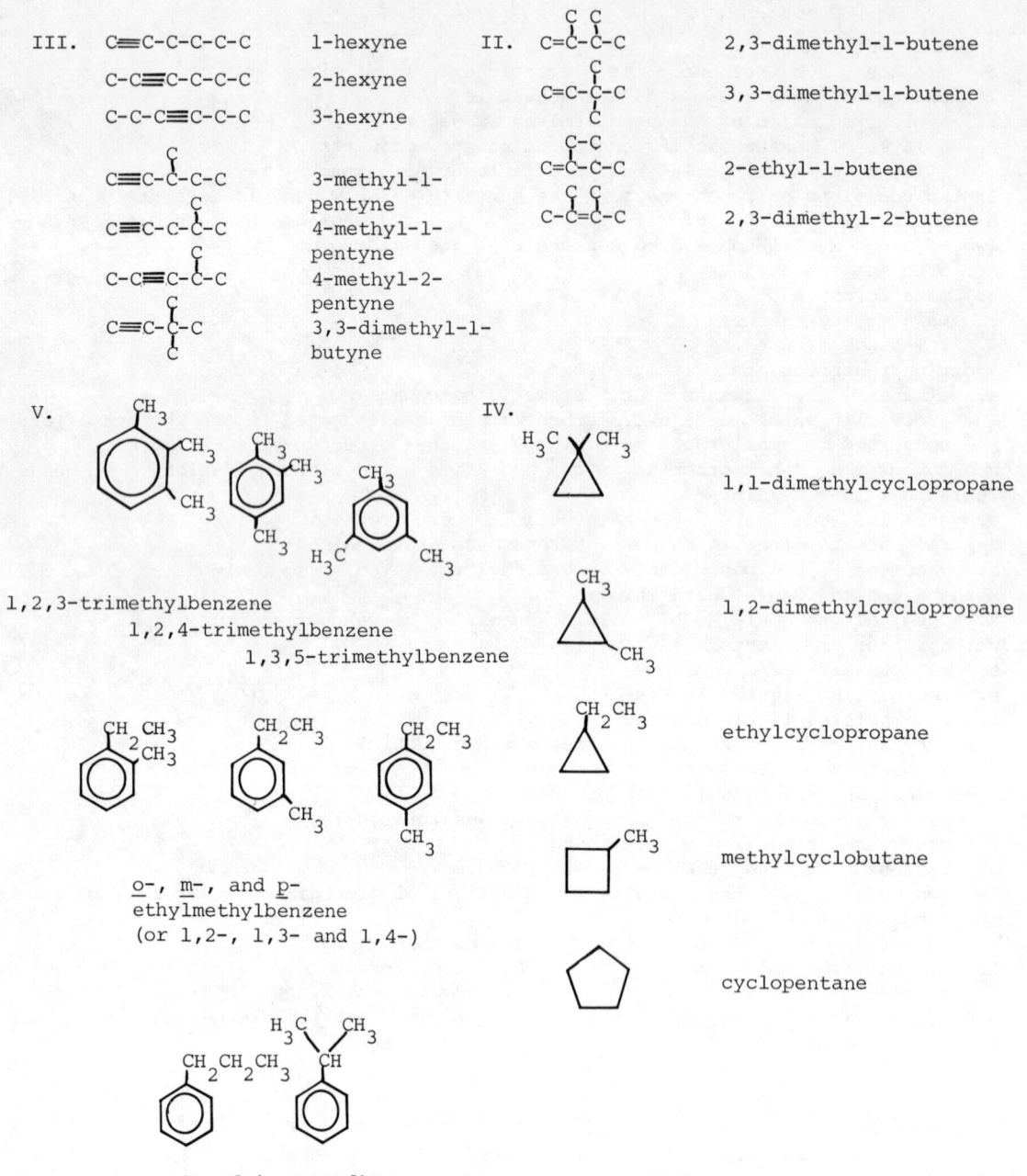

III.
C≡C-C-C-C-C   1-hexyne
C-C≡C-C-C-C   2-hexyne
C-C-C≡C-C-C   3-hexyne

C≡C-C(C)-C-C   3-methyl-1-pentyne
C≡C-C-C(C)-C   4-methyl-1-pentyne
C-C≡C-C(C)-C   4-methyl-2-pentyne
C≡C-C(C)(C)-C   3,3-dimethyl-1-butyne

II.
C=C(C)-C(C)-C   2,3-dimethyl-1-butene
C=C-C(C)(C)-C   3,3-dimethyl-1-butene
C=C(C-C)-C-C   2-ethyl-1-butene
C-C(C)=C(C)-C   2,3-dimethyl-2-butene

V.
1,2,3-trimethylbenzene
1,2,4-trimethylbenzene
1,3,5-trimethylbenzene

o-, m-, and p-ethylmethylbenzene
(or 1,2-, 1,3- and 1,4-)

n-propyl- and isopropylbenzene

IV.
1,1-dimethylcyclopropane
1,2-dimethylcyclopropane
ethylcyclopropane
methylcyclobutane
cyclopentane

Self-Test
1. c
2. b
3. b
4. a and d
5. e
6. b
7. c
8. a
9. a
10. a
11. c
12. b
13. c
14. b
15. a
16. A. b
    B. d
    C. e
17. b
18. c
19. a
20. d
21. a
22. d
23. a
24. c
25. d
26. b
27. c
28. a
29. b
30. b
31. a
32. b
33. c
34. c
35. c
36. b
37. b
38. A. a
    B. b
    C. b
    D. a
    E. a

Chapter 15 Halogenated Hydrocarbons

The questions at the end of chapter 15 serve as a thorough review of the chapter's contents. Questions 2, 3, 4 and 5 deal with nomenclature. To supplement the practice these problems offer, the following summary restates briefly the more important rules for naming halocarbons.

IUPAC--The halogens are names as substituents of a parent hydrocarbon. As substituents, the halogens are referred to by the prefixes: fluoro-, chloro-, bromo-, and iodo-.
Examples: chloromethane, 1-bromo-2-methylpropane, fluoroethene, iodocyclopentane, 1,3,5-trichlorobenzene

Common--Alkyl halide--Compounds are named as derivatives of the halides (fluoride, chloride, bromide, and iodide) with carbon groups named as substituents. Examples: methyl chloride, isobutyl bromide, vinyl fluoride, cyclopentyl iodide

Modified "IUPAC"--For the smaller halogenated alkenes, the common alkene names (ethylene, propylene, isobutylene) are used with the halogen prefixes employed in the IUPAC system. Examples: fluoroethylene, 1,2-dichloropropylene

Special--The chapter is filled with compounds referred to by special names, e.g., chloroform, freons, DDT, endrin, PCBs, hexachlorophene, etc. You should know the structure of chloroform (CHCl₃) and the general structural features of freons and PCBs. The freons are small organic molecules substituted with fluorine and, frequently, chlorine. The PCBs feature two benzene rings connected by a bond and substituted with chlorines. (Phenyl is the name given to a benzene ring when it is considered a substituent group). The rest of the compounds should be recognized by name simply as halogenated organic molecules.

Self-Test - Select the best answer.
1. Which is not an appropriate name for CH₂=CH-Cl?
    a. chloroethylene   b. chloroethene   c. methylene chloride   d. vinyl chloride
2. Which is not a freon-type compound?

    a.  F-C(F)(Cl)-F   b. H-C(Cl)(Cl)-Cl   c. F-C(Cl)(Cl)-Cl   d. 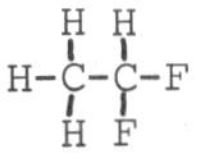 H-C(H)(H)-C(H)(F)-F

3. Which is a member of the PCB group of compounds?

    a. 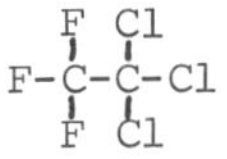 F-C(F)(F)-C(Cl)(Cl)-Cl   b. (nitrochlorobenzene)   c. (dichlorobiphenyl)   d. 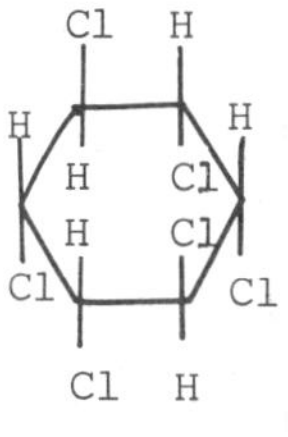 (chlorinated cyclohexane)

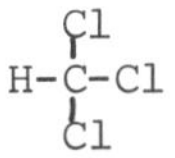

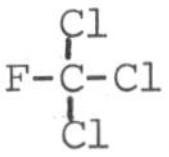

4. The IUPAC name of

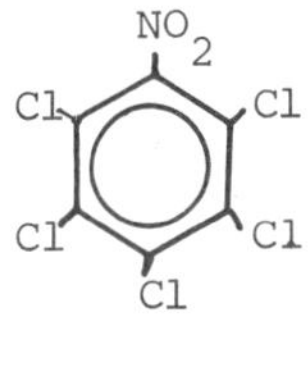

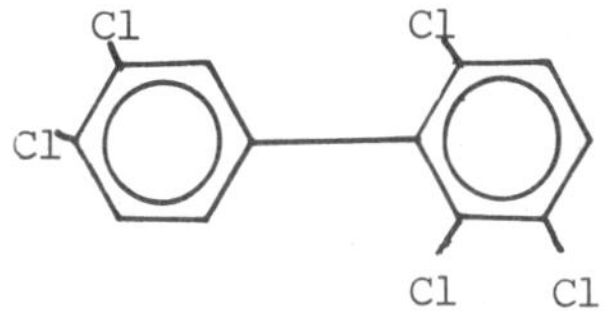

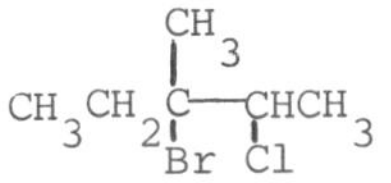

    CH₃CH₂-C(CH₃)(Br)-CH(Cl)CH₃   is

    a. 3-bromo-4-chloro-3-methylpropane
    b. 2-chloro-3-bromo-3-hexane
    c. 3-bromo-2-chloro-3-methylpentane
    d. 3-bromo-4-chloro-3-methylpentane

5. Perchlorobenzene is

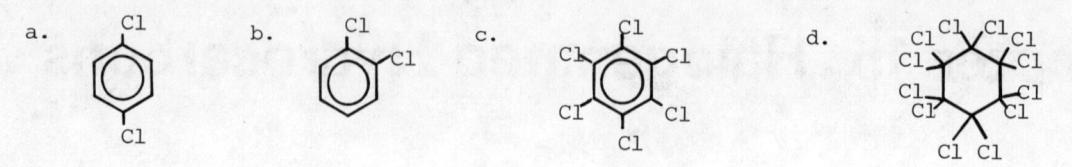

   a. [benzene with Cl at 1,4 positions]   b. [benzene with Cl at 1,2 positions]   c. [benzene with Cl at all 6 positions]   d. [cyclohexane with Cl at all 12 positions]

6. Which is <u>not</u> an alkyl halide?

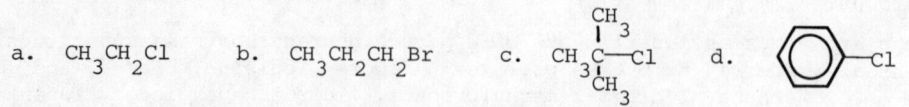

   a. CH_3CH_2Cl b. $CH_3CH_2CH_2Br$ c. $CH_3\underset{CH_3}{\overset{CH_3}{\underset{|}{\overset{|}{C}}}}Cl$ d. [phenyl-Cl]

7. For which compound is the symbol Ar-Br <u>not</u> appropriate?

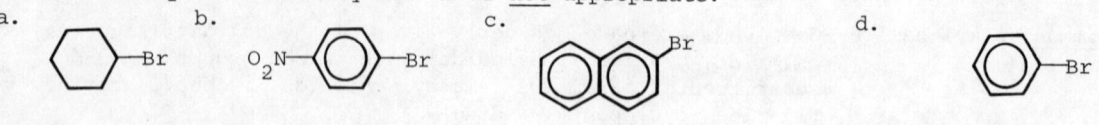

   a. [cyclohexyl-Br]   b. O_2N-[phenyl]-Br   c. [naphthyl-Br]   d. [phenyl-Br]

8. Which is <u>not</u> an aliphatic compound?

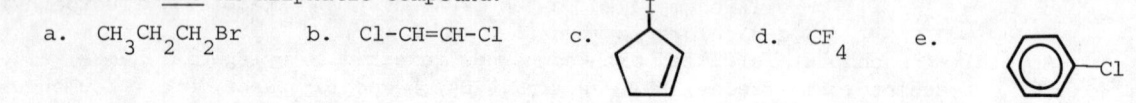

   a. $CH_3CH_2CH_2Br$ b. $Cl-CH=CH-Cl$ c. [cyclopentene with I]   d. CF_4 e. [phenyl-Cl]

9. Which is <u>not</u> a reason for discontinuing the use of carbon tetrachloride in fire extinguishers?
   a. The compound is flammable.
   b. Exposure to the compound itself can cause severe liver damage.
   c. The compound can react with water to form the extremely toxic gas phosgene.
   d. All of the above are valid reasons for discontinuing use of CCl_4 in fire extinguishers.

10. Which compound enjoyed widespread use as an anesthetic at one time?
    a. CH_3Cl b. CH_2Cl_2 c. $CHCl_3$ d. CCl_4

11. The most important use for vinyl chloride is
    a. as a dry cleaning solvent
    b. as the starting material for the synthesis of vinyl plastics
    c. as a pesticide
    d. as an anesthetic
    e. as a plasticizer

12. If lard, an animal fat, is mixed with chloroform,
    a. a clear solution of fat dissolved in chloroform is formed
    b. fat globules settle to the bottom of the chloroform layer
    c. fat globules float on top of the chloroform layer

13. If water and chloroform are mixed,
    a. the two miscible liquids form a clear solution
    b. a water layer floats on top of a chloroform layer
    c. a chloroform layer floats on top of a water layer

14. Which compound would propyl chloride most closely resemble in boiling point?
    a. propane (C_3H_8)   b. pentane (C_5H_{12})   c. heptane (C_7H_{16})

15. The freons have been used as
    a. refrigerants
    b. propellants in aerosol spray products
    c. insecticides
    d. both a and b
    e. a, b, and c

16. Potential dangers associated with the use of freons include
    a. the risk of heart failure triggered by inhalation of freon aerosol propellants
    b. an increase in the concentration of toxic ozone in the stratosphere because of

reactions involving freons
- c. an increase in the incidence of liver cancer among workers converting freons to plastics
- d. both a and b
- e. a, b and c

17. The concentration of DDT up the food chain results from its
    - a. fat-soluble nature
    - b. density
    - c. toxicity
18. The threat to bird populations posed by DDT is a result of the compound's
    - a. action as a nerve poison
    - b. interference with calcium metabolism
    - c. carcinogenic activity
19. Which is not a reason for discontinuing the use of DDT?
    - a. the development of resistant strains of insects
    - b. the persistence of DDT in the environment
    - c. the nonselectivity of DDT in attacking animal life
    - d. the proven ineffectiveness of DDT in controlling the spread of diseases like malaria
20. Which is not an insecticide?
    - a. Chlordane
    - b. Halothane
    - c. Heptachlor
    - d. Dieldrin
    - e. Endrin

ANSWERS

1. c	11. b
2. b	12. a
3. c	13. b
4. c	14. b
5. c	15. d
6. d	16. a
7. a	17. a
8. e	18. b
9. a	19. d
10. c	20. b

Chapter 16 Alcohols, Phenols, and Ethers

We'll briefly outline here some of the distinguishing characteristics of the classes of compounds discussed in chapter 16.

I. Alcohols
   A. General structure - R-OH
   B. Nomenclature
      1. Common--the carbon-containing group is specified and followed by the word alcohol. Examples are methyl alcohol and sec-butyl alcohol
      2. IUPAC--the -e ending of the parent alkane name is dropped and replaced by -ol; the position of the hydroxyl group is specified if necessary. Examples are methanol and 2-butanol.
   C. Physical properties
      1. The alcohols are relatively high boiling compared to hydrocarbons because alcohol molecules are able to hydrogen bond with one another.
      2. The alcohols are relatively water-soluble compared to hydrocarbons because of alcohol molecules' ability to form hydrogen bonds with water molecules.
   D. Reactions of interest
      1. Preparation of alcohols
         a. Hydration of alkenes: $-C=C- \xrightarrow{H^+, H_2O} -\overset{H}{\underset{|}{C}}-\overset{OH}{\underset{|}{C}}-$
         b. Fermentation (for ethanol): see section 16.4
         c. Special reaction for methanol: $CO + 2 H_2 \longrightarrow CH_3OH$
      2. Reactions of alcohols
         a. Dehydration
            1. Alkene formation: $-\overset{H}{\underset{|}{C}}-\overset{OH}{\underset{|}{C}}- \xrightarrow{\text{conc. } H_2SO_4, \text{ heat}} -C=C- + H_2O$
            2. Ether formation: $-C-O-H + H-O-C- \xrightarrow{\text{conc. } H_2SO_4, \text{ heat}} -C-O-C- + H_2O$
         b. Oxidation
            1. Primary alcohols: $-\overset{O-H}{\underset{H}{\overset{|}{C}}}-H \xrightarrow{[O]} -\overset{O}{\overset{\|}{C}}-H \xrightarrow{[O]} -\overset{O}{\overset{\|}{C}}-OH$
               aldehyde   carboxylic acid
            2. Secondary alcohols: $-\overset{O-H}{\underset{|}{C}}-H \xrightarrow{[O]} -\overset{O}{\overset{\|}{C}}-$
               ketone
            3. Tertiary alcohol: $-\overset{O-H}{\underset{|}{C}}- \xrightarrow{[O]}$ No reaction

II. Phenols
    A. General structure - Ar-OH
    B. Nomenclature--Simple compounds are named as derivatives of phenol, but many of the most interesting phenols have special common names. Examples: o-nitrophenol (a systematic name), hexachlorophene (a common name)
    C. Physical properties
       1. Compared to hydrocarbons, phenols have relatively high boiling points (most are solids at room temperature) because of intermolecular hydrogen bonding.
       2. Compared to hydrocarbons, phenols are relatively water-soluble because of interactions through hydrogen bonding with water molecules. However, by definition phenols have at least 6 carbon atoms, which is the borderline of water-solubility for compounds containing one oxygen.

III. Ethers
     A. General structure - R-O-R, R-O-Ar, Ar-O-Ar
     B. Nomenclature--Simple ethers are named by specifying the carbon-containing groups attached to the oxygen atom and adding the word ether, viz., dimethyl ether, ethyl methyl ether.

C. Physical properties
1. Ethers have boiling points comparable to hydrocarbons of similar molecular weight because ether molecules cannot hydrogen bond with one another.
2. Ethers are more soluble in water than corresponding hydrocarbons because ethers can interact with water molecules through hydrogen bonding.
D. Reactions of interest--Ethers are characterized by their relative chemical inertness. The only reaction noted was peroxide formation, considered an undesirable chemical property because of the explosive nature of peroxides.

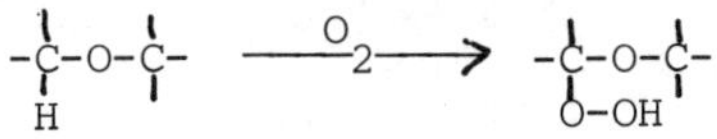

$$-\overset{|}{\underset{H}{C}}-O-\overset{|}{\underset{|}{C}}- \xrightarrow{O_2} -\overset{|}{\underset{O-OH}{C}}-O-\overset{|}{\underset{|}{C}}-$$

Questions 4, 5 and 6 at the end of the chapter offer practice in naming and drawing compounds. Questions 13, 14, 15, 16 and 18 deal with chemical reactions involving alcohols. The remaining questions in the text and the self-test here review the rest of the chapter.

Self-Test

1. A combination of atoms which confers certain chemical and physical properties on a compound is called a(n):
   a. ether   b. functional group   c. hydrogen bond

2. Which is an alcohol?
   a. C_6H_6 b. $CH_3-\overset{O}{\overset{\|}{C}}-CH_3$ c. CH_3-O-CH_3 d. C_2H_5OH

3. Which compound is an ether?
   a. OH OH OH
      | | |
      CH_2CH-CH
   b. CH_3CH_2OH c. $CH_3-O-C_3H_7$ d. $CH_3-\overset{O}{\overset{\|}{C}}-CH_3$

4. The presence of a hydroxyl group attached directly to a benzene ring makes the compound a(n)
   a. alcohol   b. ether   c. phenol   d. base   e. explosive

5. The compound [cyclic ether structure] is a(n):
   a. alcohol   b. aldehyde   c. benzene   d. ether   e. phenol

6. Which compound is a glycol?
   a. CH_3CH_2OH b. CH_3-O-CH_3 c. OH OH
                                             | |
                                             CH_2CH_2 d. $CH_3-\overset{O}{\overset{\|}{C}}-H$ e. 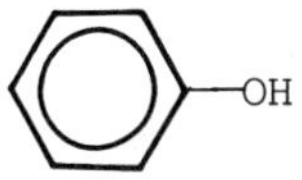 [phenol structure]

7. For which compound is R-O-Ar an appropriate abbreviation?
   a. [C₆H₅-OH]   b. CH_3OH c. CH_3-O-CH_3 d. CH_3-O-[C₆H₅]   e. [C₆H₅-O-C₆H₅]

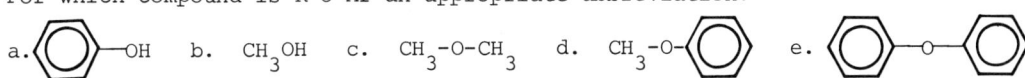

8. Which compound is a trihydric alcohol?
   a. rubbing alcohol   b. propylene glycol   c. glycerol

9. Which is a primary alcohol?
   a. CH_3OH b. CH_3CH-CH_2OH
                        |
                        CH_3 c. $CH_3CH_2CHCH_3$
                                          |
                                          OH   d. $CH_3\overset{OH}{\overset{|}{C}}CH_3$
                                                         |
                                                         CH_3

10. Which is a secondary alcohol?
    a. $CH_3CH_2CH_2CH_2OH$ b. $CH_3CH_2CH_2CHOH$
                                           |
                                           CH_3 c. $CH_3\overset{CH_3}{\overset{|}{C}}-OH$
                                                              |
                                                              CH_3 d. $CH_3-O-CH_2CH_3$ e. $CH_3\overset{OH}{\overset{|}{C}}H-OH$

11. Which compound is a primary alcohol?
    a. [phenol]   b. [cyclohexanol]   c. [cyclohexyl-CH₂OH]   d. [1-methylcyclohexanol]

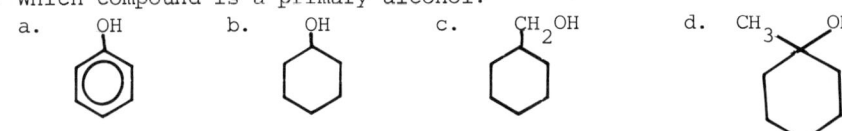

12. The compound $CH_3CH_2\overset{}{\underset{CH_3}{C}H-OH}$ is not properly called
    a. 2-butanol   b. isobutyl alcohol   c. sec-butyl alcohol

13. The correct name of CH_3-O-CH_3 is
    a. dimethyl ether   b. ethyl ether   c. diethyl ether   d. dimethyl oxide

61

14. Wood alcohol is the same as
    a. methanol b. 2-propanol c. glycerin d. grain alcohol e. rubbing alcohol
15. Phenol is
    a. [cyclohexanol structure] OH   b. [phenol structure] OH   c. $CH_3CH_2CH_2-OH$ d. $CH_3CH_2CH_2CH_2CH_2OH$
16. The formula of anesthetic "ether" is
    a. CH_3OCH_3 b. C_2H_5OH c. $CH_3CH_2\overset{O}{\overset{\|}{C}}CH_2CH_3$ d. $C_2H_5-O-C_2H_5$
17. The alcohol present in alcoholic beverages is
    a. methyl alcohol b. ethyl alcohol c. denatured alcohol d. wood alcohol
18. Alcohols boil at appreciably higher temperatures than hydrocarbons of similar formula weight because
    a. they have a higher molecular weight
    b. alcohols are strongly acidic
    c. alcohols are ionic compounds
    d. the molecules of the alcohol are associated through hydrogen bonding
    e. alcohols are soluble in water
19. Which compound would have the highest boiling point?
    a. $CH_3CH_2-O-CH_2CH_3$ b. $CH_3OCH_2CH_2OH$ c. $CH_3\underset{OH}{C}H-\underset{OH}{C}H_2$ d. $CH_3CH_2CH_2CH_2OH$
20. Which compound would be expected to have the lowest boiling point?
    a. $CH_3CH_2CH_2OH$ b. $CH_3-O-CH_2CH_3$ c. $HO-CH_2CH_2OH$
21. Which alcohol is least soluble in water?
    a. CH_3OH b. C_3H_7OH c. $C_6H_{13}OH$ d. $C_{10}H_{21}OH$
22. Which compound would be most soluble in water?
    a. $CH_2=CHCH_2CH_3$ b. $CH_3CH_2OCH_2CH_3$ c. $CH_3CH_2\underset{OH}{C}H-\underset{OH}{C}H_2$
23. Which compound is not miscible with water?
    a. CH_3CH_2OH b. $\underset{OH}{C}H_2\underset{OH}{C}H_2$ c. [phenol structure with OH]

24. Fermentation of carbohydrates leads to the formation of
    a. methyl alcohol b. ethyl alcohol c. glucose d. 1-propanol e. formaldehyde
25. Which solution could not be obtained directly from the fermentation reaction?
    a. 10% ethyl alcohol in water
    b. 30% ethyl alcohol in water
    c. 10 proof ethyl alcohol
    d. 30 proof ethyl alcohol
26. Dehydration of alcohols does not produce
    a. ethers   b. alkenes   c. aldehydes
27. Which term is used to describe the reaction: $CH_2=CH_2 + H_2O \xrightarrow{H^+} \underset{OH}{C}H_2-\underset{H}{C}H_2$?
    a. combustion   b. dehydration   c. hydration   d. oxidation
28. Markovnikov's rule indicates that
    a. when alcohols are dehydrated, they yield alkenes in preference to ethers
    b. primary and secondary alcohols are readily oxidized, but tertiary alcohols are not
    c. when water adds to alkenes, the hydrogen of water adds to the double-bonded carbon with most hydrogens
29. Which product would be expected form the hydration of 1-butene?
    a. $CH_3CH_2CH_2CH_2OH$ b. $CH_3\underset{OH}{C}H-CH_2CH_3$ c. $CH_3CH_2CH_2-O-CH_3$ d. $CH_3\underset{CH_3}{\overset{OH}{C}}-CH_3$
30. Which compound is not readily oxidized?
    a. $CH_3\underset{CH_3}{C}H-CH_2OH$ b. $CH_3CH_2\underset{OH}{C}H-CH_3$ c. $CH_3\underset{CH_3}{\overset{OH}{C}}-CH_3$ d. $CH_3CH_2\underset{H}{C}=O$

31. What is the product of the reaction: $CH_3\underset{}{\overset{OH}{\underset{|}{C}H}}-CH_3 \xrightarrow{K_2Cr_2O_7,\ H^+}$?
    a. CH_3COOH b. $CH_3CH_2CH_2OH$ c. $CH_3CH_2\underset{H}{\overset{}{C}}=O$ d. CH_3CH_2COOH e. $CH_3\underset{O}{\overset{\|}{C}}-CH_3$

32. In chemical reactivity, ethers resemble
    a. alcohols   b. alkanes   c. phenols

33. Air oxidation of ethers results in the formation of
    a. aldehydes   b. carboxylic acids   c. ketones   d. peroxides

34. Denatured alcohol refers to
    a. any alcohol not produced by fermentation
    b. grain alcohol which is highly taxed
    c. ethyl alcohol which has been treated with something to make it unfit to **drink**

35. The toxicity of wood alcohol results from its oxidation by liver enzymes to
    a. carbon dioxide   b. formaldehyde   c. grain alcohol   d. methanol

36. When ingested, ethyl alcohol's physiological action is that of a(n):
    a. stimulant   b. depressant   c. antiseptic

37. Antiseptic properties are <u>not</u> generally attributed to
    a. the smaller alcohols
    b. a number of phenol derivatives
    c. diethyl ether

38. Which is quite toxic when ingested by humans?
    a. ethylene glycol   b. propylene glycol   c. glycerol

39. Hexachlorophene, a germicide once widely used, is <u>not</u> a(n):
    a. ether   b. halogenated aromatic compound   c. phenol

40. Which compound is widely used as a general anesthetic?
    a. dimethyl ether   b. diethyl ether   c. bis(chloromethyl)ether (BCME)

ANSWERS

1. b	11. c	21. d	31. e
2. d	12. b	22. c	32. b
3. c	13. a	23. c	33. d
4. c	14. a	24. b	34. c
5. d	15. b	25. b	35. b
6. c	16. d	26. c	36. b
7. d	17. b	27. c	37. c
8. c	18. d	28. c	38. a
9. b	19. c	29. b	39. a
10. b	20. b	30. c	40. b

Chapter 17 Aldehydes and Ketones

Figures 17.1 and 17.2 in the chapter illustrate the nomenclature rules discussed in sections 17.2 and 17.3. You can use these figures to check your understanding of the common and IUPAC naming systems. If you have not already done so, answer the first three questions at the end of chapter 17 for added practice.

By far, the greater part of this chapter was devoted to chemical reactions involving aldehydes and ketones. The selection of reactions considered was based on their significance in living systems. It will not be long before we encounter aldol concentrations, keto-enol tautomerism, and acetal formation in our study of the chemistry of carbohydrates. Carbohydrates contain many functional groups and, on first encounter, strike one as very complex molecules. That is why we are taking the time now to look at reactions which we'll consider again later. By looking at simple molecules, we can concentrate on the general pattern of these reactions. That same pattern will be followed when the reacting molecules are more complex.

The reactions of chapter 17 are gathered below for easy reference.

Summary of Reactions of Aldehydes and Ketones

I. Oxidation - This is the reaction which most clearly distinguishes the aldehyde family from the ketone family.

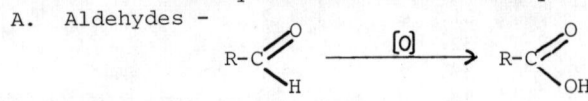

   A. Aldehydes -

   $$R-CHO \xrightarrow{[O]} R-COOH$$

   The usual oxidizing agents (like $K_2Cr_2O_7$) work, as do much weaker oxidizing agents (like Ag^+ in the form of Tollens' reagent)

   B. Ketones - Under ordinary conditions, ketones give no reaction.

II. Reactions involving addition to the carbonyl group

   A. Addition of water (hydrate formation)

   $$-\overset{O}{\underset{\|}{C}}- \quad H\text{-}O\text{-}H \longrightarrow -\overset{O-H}{\underset{|}{C}}-O\text{-}H$$

   B. Addition of alcohol (hemiacetal and acetal formation)
      1. Hemiacetal formation

      $$-\overset{O}{\underset{\|}{C}}- \quad H\text{-}O\text{-}R \longrightarrow -\overset{O-H}{\underset{|}{C}}-OR$$

      2. Acetal formation (accompanied by elimination of water)

      $$-\overset{H\ \ OH}{\underset{|}{C}}\overset{R}{\underset{OR}{}} \xrightarrow{dry\ HCl} -\overset{H\ \ \ OR}{\underset{|}{C}}\overset{R}{\underset{OR}{}}$$

      3. Shortcut for determining acetal formed from one carbonyl and 2-alcohol units

      $$\overset{H\text{-}OR}{\underset{H\text{-}OR}{}}C=O \longrightarrow \overset{OR}{\underset{OR}{}}C \quad H_2O$$

   C. Addition of ammonia and derivatives
      1. Imine formation (accompanied by elimination of water)

      $$-\overset{O}{\underset{\|}{C}}- \quad H\text{-}N\text{-}H \longrightarrow \left[-\overset{O-H}{\underset{|}{C}}-N\text{-}H\right] \longrightarrow -C=N\text{-}H \quad H_2O$$

      2. Phenylhydrazone formation (accompanied by elimination of water)

      $$-\overset{O}{\underset{\|}{C}}- \quad H\text{-}N\text{-}NHC_6H_5 \longrightarrow \left[-\overset{O-H}{\underset{|}{C}}-N\text{-}NHC_6H_5\right] \longrightarrow -C=N\text{-}NHC_6H_5 \quad H_2O$$

3. Shortcut for determining product formed from ammonia or derivative

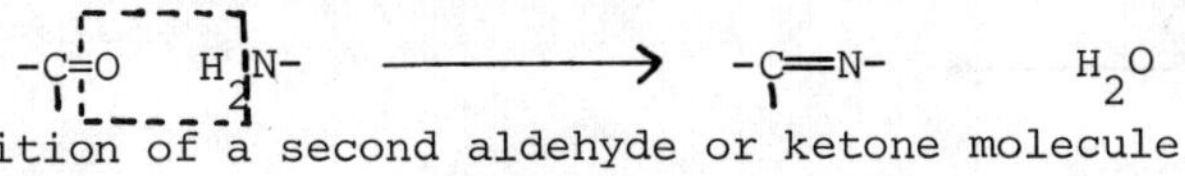

$$-\overset{|}{\underset{|}{C}}=O \quad H_2N- \longrightarrow \quad -\overset{|}{\underset{|}{C}}=N- \quad H_2O$$

D. Addition of a second aldehyde or ketone molecule (aldol condensation)

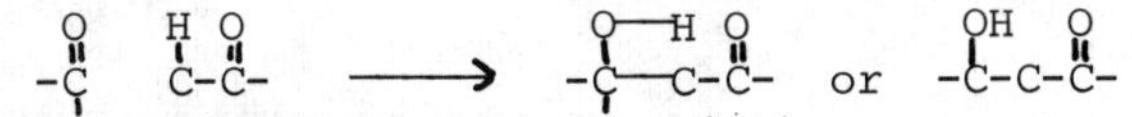

$$-\overset{O}{\underset{}{C}}-\overset{H}{\underset{}{C}}-\overset{O}{\underset{}{C}}- \longrightarrow -\overset{O-H}{\underset{}{C}}-\overset{}{\underset{}{C}}-\overset{O}{\underset{}{C}}- \quad or \quad -\overset{OH}{\underset{}{C}}-\overset{}{\underset{}{C}}=\overset{O}{\underset{}{C}}-$$

III. Isomerism (Keto-enol tautomerism)

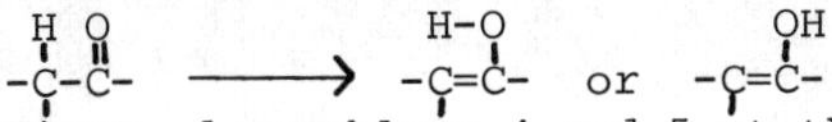

$$-\overset{H}{\underset{}{C}}-\overset{O}{\underset{}{C}}- \longrightarrow -\overset{H-O}{\underset{}{C}}=\overset{}{\underset{}{C}}- \quad or \quad -\overset{OH}{\underset{}{C}}=\overset{}{\underset{}{C}}-$$

First, do problems 4 and 5 at the end of the chapter in the text, then try the following.

Draw the organic products of the reactions shown below. Some of the products of these reactions may look unusual. Just remember that the reactions follow exactly the same pattern as those shown above.

A. 1. $C_6H_5-\underset{H}{\overset{|}{C}}=O + CH_3OH \longrightarrow$

2. $C_6H_5-\underset{H}{\overset{|}{C}}=O + 2\ CH_3CH_2OH \xrightarrow{dry\ HCl}$

3. $C_6H_5-\underset{H}{\overset{|}{C}}=O + \begin{matrix}HO-CH_2\\ HO-CH_2\end{matrix} \xrightarrow{dry\ HCl}$

4. cyclohexanone $+ 2\ CH_3OH \xrightarrow{dry\ HCl}$

5. $H-\underset{O}{\overset{}{C}}-CH_2CH_2\overset{OH}{\underset{}{CH_2}} \longrightarrow$ a hemiacetal

B. 1. $C_6H_5-\underset{H}{\overset{|}{C}}=O \xrightarrow{K_2Cr_2O_7,\ H^+}$

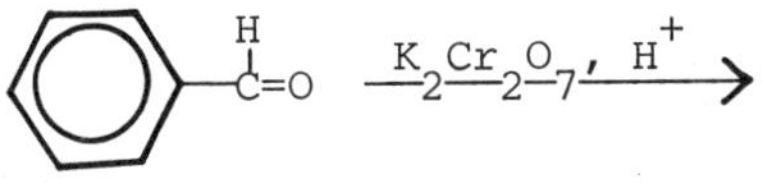

2. cyclohexanone $\xrightarrow{K_2Cr_2O_7,\ H^+}$

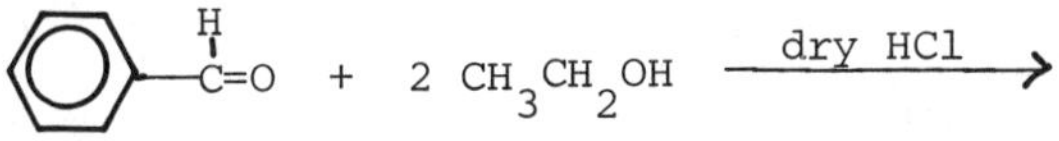

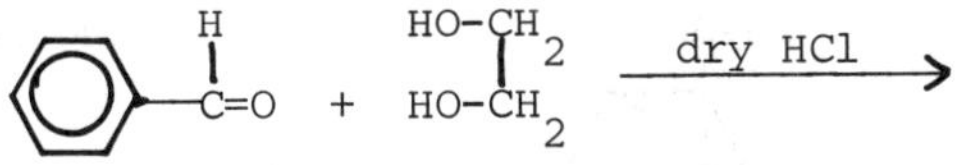

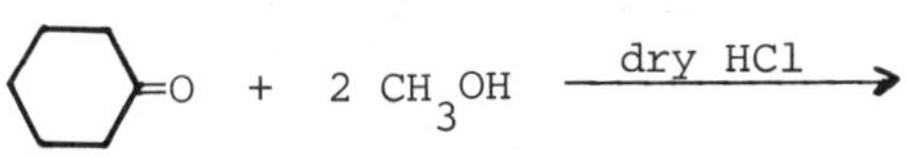

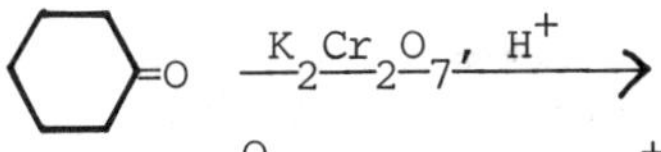

3. $CH_3CH_2CH_2\overset{O}{\overset{\|}{C}}CH_3 \xrightarrow{Ag(NH_3)_2^+\ OH^-}$

4. $CH_3-\underset{CH_3}{\overset{CH_3}{\underset{|}{\overset{|}{C}}}}-\underset{H}{\overset{}{C}}=O \xrightarrow{Ag(NH_3)_2^+\ OH^-}$

5. $CH_3-\overset{O}{\overset{\|}{C}}-CH_2CH_2\underset{H}{\overset{}{C}}=O \xrightarrow{K_2Cr_2O_7,\ H^+}$

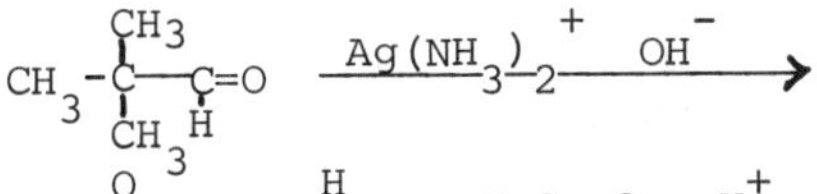

C. 1. $C_6H_5-\underset{H}{\overset{|}{C}}=O \xrightarrow{NH_3}$

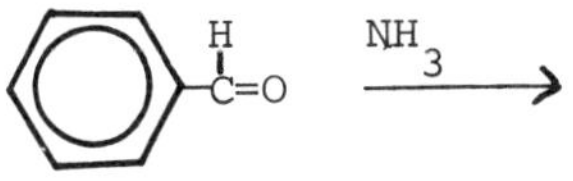

2. $C_6H_5-\underset{H}{\overset{|}{C}}=O \xrightarrow{H_2N-NH-C_6H_5}$

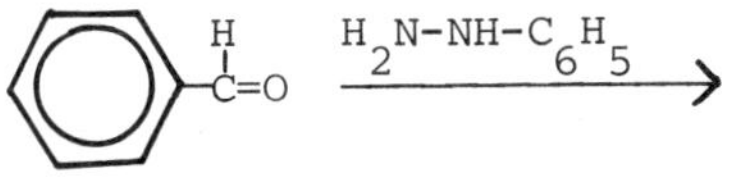

3. cyclohexanone + H₂N-NH-C₆H₅ →

4. CH₃-C(=O)-CH₂CH₂-C(=O)-H + H₂N-NHC₆H₅ →

5. CH₃-C(=O)-COOH + NH₃ →

Self-Test

1. A group which both aldehydes and ketones have in common is
   a. -COOH   b. -C(H)=O   c. -C=O   d. -OH   e. -O-
2. The name of the functional group of aldehydes and ketones is
   a. carbonyl group   b. carboxyl group   c. double bond   d. hydroxyl group
3. Which structural feature is possessed by aldehydes but not ketones?
   a. an alpha hydrogen
   b. a hydrogen on the carbonyl carbon
   c. a hydroxyl group on the carbonyl carbon
4. The compound H₂C=O is
   a. methyl alcohol   b. formic acid   c. formaldehyde   d. acetone   e. methanone
5. Benzaldehyde is
   a. CH₃CH₂CH₂CH=O   b. (C₆H₅)₂C=O   c. CH₃-C(=O)-C₆H₅   d. C₆H₅-C(=O)-H
6. The name of CH₃CH₂CH=O is not
   a. propanal   b. propanone   c. propionaldehyde
7. The name of CH₃CH₂CH₂C(=O)CH₃ is
   a. butyl methyl ketone   b. 2-pentanone   c. 4-pentanone   d. 2-pentyl aldehyde
8. CH₃-C(=O)-CH₃ is not
   a. acetone   b. dimethyl aldehyde   c. dimethyl ketone   d. propanone
9. The compound 3-chlorobutanal is also properly called
   a. 3-chlorobutanol   b. 3-chlorobutyraldehyde   c. α-chlorobutanal
   d. α-chlorobutyraldehyde   e. β-chlorobutyraldehyde
10. An aqueous solution of formaldehyde is called
    a. aldol   b. acetone   c. formalin   d. formic acid   e. methanal
11. In general, aldehydes and ketones exhibit greater water-solubility than
    a. alcohols   b. alkenes   c. ethers
12. Which compound has the lowest boiling point?
    a. CH₃CH₂OH   b. CH₃CH=O   c. CH₃-O-CH₃
13. Tollens' reagent will oxidize a(n):
    a. aldehyde   b. alcohol   c. ketone   d. carboxylic acid
14. When an aldehyde is oxidized, the product is a(n):
    a. alcohol   b. ketone   c. carboxylic acid   d. aldehyde
15. Which structure would give a positive Tollens' test?
    a. CH₃CH=O   b. CH₃-C(=O)-CH₃   c. cyclopentanone   d. CH₃COOH
16. An aldehyde can be distinguished from a ketone by means of
    a. the Tollens' test
    b. reaction with phenylhydrazine
    c. Both contain carbonyl groups and cannot be distinguished by the above tests.
17. A ketone can be distinguished from an alcohol by means of
    a. the Tollens' test
    b. reaction with phenylhydrazine
    c. Both contain oxygen and cannot be distinguished by the above tests.

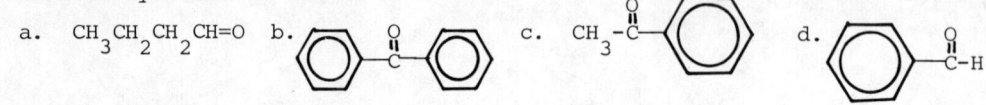

18. Ketones are prepared by the oxidation of
    a. primary alcohols
    b. secondary alcohols
    c. tertiary alcohols
    d. carboxylic acids
    e. Ketones can not be prepared by oxidation reactions.
19. Which reaction does not involve addition to the aldehyde carbonyl group?
    a. hydrate formation   b. hemiacetal formation   c. imine formation   d. oxidation
20. Both tautomerism and the aldol condensation depend on the relative acidity of the
    a. hydrogen alpha to the carbonyl group
    b. hydrogen on the carbonyl carbon atom
    c. carbon of the carbonyl group
    d. oxygen of the carbonyl group
21. In general, which is the most stable type of compound?
    a. hydrate   b. hemiacetal   c. acetal
22. Which organic compound would be isolated from the reaction: $CH_3\overset{O}{C}CH_3 \xrightarrow{K_2Cr_2O_7, H^+}$?
    a. CH_3CH_2COOH b. $CH_3\overset{COOH}{\underset{}{C}H}CH_3$ c. $CH_3\overset{O}{C}CH_3$ d. $CH_3\overset{O}{C}COOH$.
23. What must be added to acetaldehyde to form $CH_3-\overset{H}{\underset{OCH_3}{C}}-OH$?
    a. CH_3CH_2OH b. $CH_3-\overset{O}{C}-H$ c. CH_3OH d. $CH_2=O$
24. Which compound represents the tautomer of $CH_3-\overset{O}{\underset{}{C}}-CH_3$?
    a. $CH_3CH_2\overset{O}{C}-H$ b. $CH_3\overset{OH}{\underset{}{C}H}=CH$ c. $CH_3\overset{OH}{\underset{}{C}H}CH_3$ d. $CH_2=\overset{OH}{\underset{}{C}}-CH_3$
25. Which of the following reagents is required to accomplish each of the transformations shown below:

    a. H^+, H_2O b. dry HCl   c. NaOH   d. $K_2Cr_2O_7$, H^+ e. no additional reagent required

    A. $CH_3\overset{O}{C}-H \longrightarrow CH_3\overset{O}{C}-OH$

       a   b   c   d   e

    B. $CH_3\overset{O}{C}-H + 2\ CH_3OH \longrightarrow CH_3-\overset{OCH_3}{\underset{H}{C}}-OCH_3$

       a   b   c   d   e

    C. $CH_3\overset{O}{C}-H + H_2O \longrightarrow CH_3-\overset{OH}{\underset{H}{C}}-OH$

       a   b   c   d   e

    D. $2\ CH_3\overset{O}{C}-H \longrightarrow CH_3-\overset{OH}{\underset{}{C}H}-CH_2-\overset{O}{C}-H$

       a   b   c   d   e

26. Oxidation of wood alcohol by liver enzymes produces the toxic substance
    a. ethanol   b. formaldehyde   c. methanol   d. nicotine
27. Which does not contain a carbonyl group?
    a. testosterone   b. progesterone   c. chloral hydrate

ANSWERS

Reactions of Aldehydes and Ketones

A. 1. $C_6H_5-\overset{OCH_3}{\underset{H}{C}}-OH$;   2. 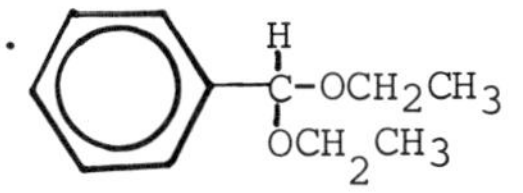 Ph–C(H)(OCH₂CH₃)(OCH₂CH₃);   3. 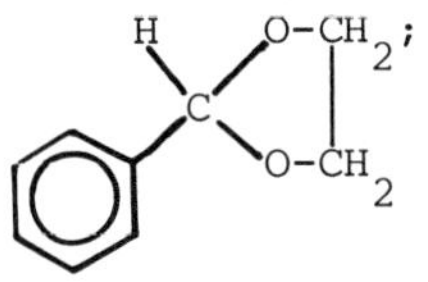 Ph-C with O-CH₂/O-CH₂ (dioxolane);   4. 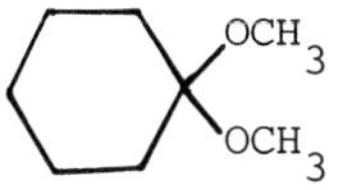 cyclohexane with (OCH₃)(OCH₃)

5. 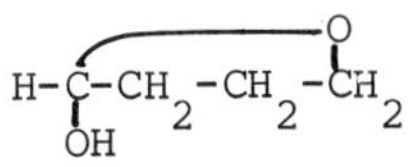 $H-\overset{}{\underset{OH}{C}}H-CH_2-CH_2-\overset{O}{C}H_2$ or   tetrahydrofuran-2-ol

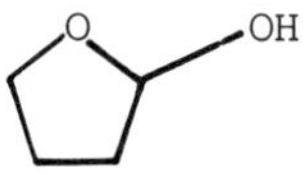

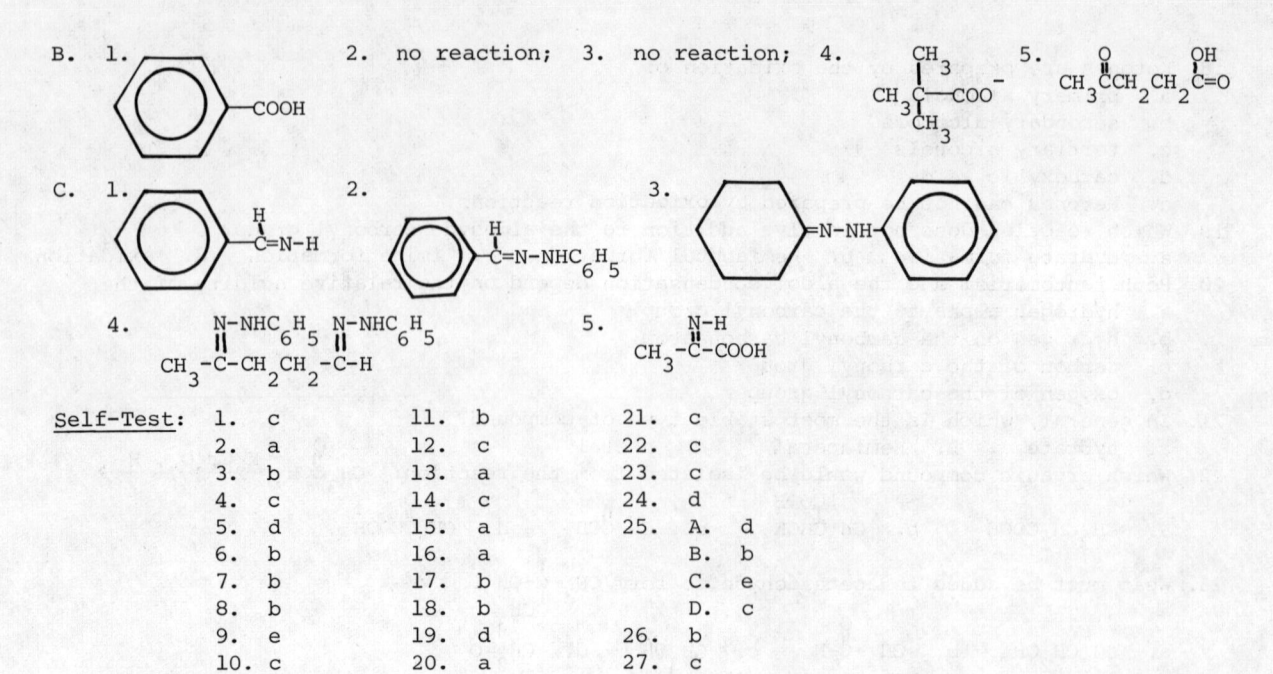

B. 1. C₆H₅-COOH 2. no reaction; 3. no reaction; 4. (CH₃)₂CH-COO⁻ 5. CH₃COCH₂CH₂COOH

C. 1. C₆H₅-CH=N-H 2. C₆H₅-CH=N-NHC₆H₅ 3. cyclohexyl=N-NH-C₆H₅

4. CH₃-C(=N-NHC₆H₅)-CH₂-C(=N-NHC₆H₅)-H   5. CH₃-C(=N-H)-COOH

Self-Test:

1. c	11. b	21. c
2. a	12. c	22. c
3. b	13. a	23. c
4. c	14. c	24. d
5. d	15. a	25. A. d
6. b	16. a	B. b
7. b	17. b	C. e
8. b	18. b	D. c
9. e	19. d	26. b
10. c	20. a	27. c

Chapter 18 Organic Acids and Derivatives

To start off, let's just summarize all of the nomenclature rules and chemical properties (by examples) covered in chapter 18.

Nomenclature of Carboxylic Acids and Derivatives

Type of Compound	Common Name	IUPAC Name	Structure
carboxylic acids	valeric acid	pentanoic acid	$CH_3CH_2CH_2CH_2\overset{O}{\overset{\|}{C}}OH$
salts of acids	sodium valerate	sodium pentanoate	$CH_3CH_2CH_2CH_2\overset{O}{\overset{\|}{C}}O^-Na^+$
esters	ethyl valerate	ethyl pentanoate	$CH_3CH_2CH_2CH_2\overset{O}{\overset{\|}{C}}OCH_2CH_3$
simple amides	valeramide	pentanamide	$CH_3CH_2CH_2CH_2\overset{O}{\overset{\|}{C}}NH_2$
substituted amides	N,N-dimethylvaleramide	N,N-dimethylpentanamide	$CH_3CH_2CH_2CH_2\overset{O}{\overset{\|}{C}}N(CH_3)_2$
anilides	valeranilide	pentananilide or N-phenylpentanamide	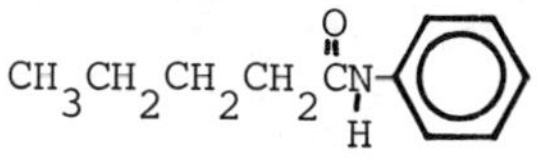 $CH_3CH_2CH_2CH_2\overset{O}{\overset{\|}{C}}NH{-}C_6H_5$

The first five questions at the end of the chapter in the text are excellent tests of your understanding of the nomenclature rules.

Reactions of Carboxylic Acids and Derivatives

Preparation of Acids (oxidation of primary alcohols or aldehydes)

$$RCH_2OH \xrightarrow{K_2Cr_2O_7,\ H^+} RCH=O \longrightarrow RCOOH$$

Have you noticed that this reaction keeps popping up. It was first discussed as a reaction of alcohols (chapter 16), then as a reaction of aldehydes (chapter 17), and now as a preparation of acids

Salt formation

$$RCOOH + NaOH \longrightarrow RCOO^-Na^+ + H_2O$$
$$RCOOH + NaHCO_3 \longrightarrow RCOO^-Na^+ + H_2O + CO_2$$
$$2\ RCOOH + Na_2CO_3 \longrightarrow 2\ RCOO^-Na^+ + H_2O + CO_2$$

Ester formation

$$R\overset{O}{\overset{\|}{C}}{-}OH + HOR' \xrightarrow{H^+} R{-}\overset{O}{\overset{\|}{C}}{-}O{-}R' + H_2O$$

$$CH_3\overset{O}{\overset{\|}{C}}{-}O{-}\overset{O}{\overset{\|}{C}}CH_3 + HOR \longrightarrow CH_3\overset{O}{\overset{\|}{C}}{-}O{-}R + CH_3\overset{O}{\overset{\|}{C}}{-}OH$$
$$\text{ester} \qquad \text{acid byproduct}$$

Ester hydrolysis

Acid-catalyzed: $R\text{-}\underset{\underset{O}{\|}}{C}\text{-}O\text{-}R' + H_2O \xrightarrow{H^+} R\text{-}\underset{\underset{O}{\|}}{C}\text{-}OH + HOR'$

Base-catalyzed: $R\text{-}\underset{\underset{O}{\|}}{C}\text{-}O\text{-}R' + H_2O \xrightarrow{OH^-} R\text{-}\underset{\underset{O}{\|}}{C}\text{-}O^- + HOR'$

Amide formation

Simple amide: $R\text{-}\underset{\underset{O}{\|}}{C}\text{-}OH \xrightarrow{SOCl_2} R\text{-}\underset{\underset{O}{\|}}{C}\text{-}Cl \xrightarrow{NH_3} R\text{-}\underset{\underset{O}{\|}}{C}\text{-}NH_2$

Substituted amide: $R\text{-}\underset{\underset{O}{\|}}{C}\text{-}OH \xrightarrow{SOCl_2} R\text{-}\underset{\underset{O}{\|}}{C}\text{-}Cl \xrightarrow{H_2N\text{-}R'} R\text{-}\underset{\underset{O}{\|}}{C}\text{-}\underset{\underset{H}{|}}{N}\text{-}R'$

$R\text{-}\underset{\underset{O}{\|}}{C}\text{-}OH \xrightarrow{SOCl_2} R\text{-}\underset{\underset{O}{\|}}{C}\text{-}Cl \xrightarrow{HNR'_2} R\text{-}\underset{\underset{O}{\|}}{C}\text{-}\underset{\underset{R'}{|}}{N}\text{-}R'$

Amide hydrolysis

Acid-catalyzed: $R\text{-}\underset{\underset{O}{\|}}{C}\text{-}NH_2 \xrightarrow{H^+, H_2O} R\text{-}\underset{\underset{O}{\|}}{C}\text{-}OH + NH_4^+$

Base-catalyzed: $R\text{-}\underset{\underset{O}{\|}}{C}\text{-}NH_2 \xrightarrow{OH^-, H_2O} R\text{-}\underset{\underset{O}{\|}}{C}\text{-}O^- + NH_3$

(substituted amides give the corresponding substituted ammonia derivatives, e.g., RNH_2 or RNH_3^+)

The following problems will test your understanding of the chemistry of acids and acid derivatives. Remember -- no matter how complicated individual structures may appear, the chemistry is determined by the functional group and follows the patterns summarized above.

Reactions - Draw the principal organic products of the reactions.

1. $\text{Ph-}\underset{\underset{O}{\|}}{C}\text{-}OH + CH_3\underset{\underset{OH}{|}}{C}HCH_3 \xrightarrow{H^+}$

2. $CH_3\underset{\underset{O}{\|}}{C}\text{-}O\text{-}\underset{\underset{O}{\|}}{C}CH_3 + \text{Ph-OH} \longrightarrow$

3. $CH_3\text{-}\underset{\underset{O}{\|}}{C}\text{-}O\text{-}\underset{\underset{O}{\|}}{C}\text{-}CH_3 +$ (testosterone structure) $\longrightarrow$

4. $HO\text{-}\underset{\underset{O}{\|}}{C}CH_2CH_2\underset{\underset{O}{\|}}{C}\text{-}OH + 2\ CH_3OH \xrightarrow{H^+}$

5. $CH_3\underset{\underset{O}{\|}}{C}\text{-}Cl + \text{Ph-}NH_2 \longrightarrow$

6. $\text{Ph-}\underset{\underset{O}{\|}}{C}\text{-}Cl + NH_3 \longrightarrow$

7. $CH_3\text{-}\underset{\underset{CH_3}{|}}{C}H\underset{\underset{O}{\|}}{C}\text{-}Cl + CH_3\text{-}\underset{\underset{H}{|}}{N}\text{-}CH_3 \longrightarrow$

8. $CH_3CH_2\underset{\underset{O}{\|}}{C}\text{-}OH + KOH \longrightarrow$

9.

   [benzene ring with two COOH groups (ortho)] + Na$_2$CO$_3$ $\longrightarrow$

10. CH$_3$CH$_2$C(=O)-O-CH$_2$CH$_3$ + H$_2$O $\xrightarrow{OH^-}$

11. CH$_3$CH$_2$C(=O)-O-CH$_2$CH$_3$ + H$_2$O $\xrightarrow{H^+}$

12. HO-[benzene ring]-NH-C(=O)-CH$_3$ $\xrightarrow{H^+, H_2O}$

13. [pyridine ring]-C(=O)-NH$_2$ $\xrightarrow{OH^-, H_2O}$

Self-Test

1. Which group bestows acidic properties on the compound to which it is attached?
   a. -CH=O   b. -NH$_2$ c. C$_6$H$_5$-   d. -CH$_2$OH   e. -C(=O)-NH$_2$ f. -C(=O)-OH

2. The -COOH group is called a(n):
   a. carboxyl group   b. carbonyl group   c. aldehyde group   d. hydroxyl group

3. Which is not a mineral acid?
   a. HNO$_3$ b. HCl   c. HCOOH   d. H$_2$SO$_4$

4. Which is the acid found in vinegar?
   a. formic acid   b. nitric acid   c. valeric acid   d. acetic acid

5. The product of the reaction between an alcohol and an acid is known as a(n):
   a. acid anhydride   b. amide   c. ester   d. ether   e. salt

6. The correct name of the compound CH$_3$CH$_2$CH$_2$CH$_2$C(=O)-OH is
   a. butanoic acid   b. caproic acid   c. succinic acid   d. valeric acid

7. CH$_3$CH$_2$CH(CH$_3$)-C(=O)-OH is
   a. α-methylbutyric acid   b. β-methylbutyric acid
   c. 2-pentanoic acid   d. 1-methylbutanoic acid

8. Ammonium acetate is
   a. CH$_3$C(=O)-O$^-$ NH$_4^+$ b. CH$_3$C(=O)-NH$_2$ c. CH$_3$C(=O)-O$^-$ Na$^+$ d. NH$_4^+$Cl$^-$

9. The name of H-C(=O)-O$^-$ Na$^+$ is not
   a. sodium carbonate   b. sodium formate   c. sodium methanoate

10. The correct name of CH$_3$CH$_2$O-C(=O)-CH$_3$ is
    a. ethyl acetate   b. ethyl formate   c. ethyl methyl ketone
    d. methyl acetate   e. ethyl methyl ester

11. Methyl acetate is
    a. C$_2$H$_5$COOCH$_3$ b. CH$_3$COOC$_2$H$_5$ c. CH$_3$COOCH$_3$ d. C$_2$H$_5$COOC$_2$H$_5$

12. Butyramide is
    a. CH$_3$CH$_2$CH$_2$C(=O)-O$^-$NH$_4^+$ b. CH$_2$(NH$_2$)CH$_2$CH$_2$C(=O)-OH   c. CH$_3$CH$_2$CH$_2$C(=O)-NH$_2$

13. CH$_3$C(=O)-NHCH$_3$ is
    a. N-methylamide   b. methylformamide   c. N-methylethanamide   d. N-methylmethanamide

14. [benzene ring]-NH-C(=O)-CH$_3$ is
    a. methylbenzamide   b. N-methylethanamide   c. methananilide   d. acetanilide

15. Glacial acetic acid is
    a. a frozen solution of acetic acid
    b. a mixture of acetic acid and water
    c. pure acetic acid
16. Which is the strongest acid?
    a. hydrochloric acid   b. benzoic acid   c. carbonic acid
17. Which compound will produce bubbles of carbon dioxide gas when mixed with sodium bicarbonate?

    a. C₆H₅-OH   b. C₆H₅-COOH   c. NaOH   d. H₂O

18. Which compound is not a stronger acid than water?
    a. CH₃CH₂OH   b. C₆H₅-OH   c. CH₃C(=O)-OH

19. Which compound is oxalic acid?
    a. C₆H₅-COOH   b. HOOC-COOH   c. HOOC-CH₂-COOH   d. benzene-1,2-dicarboxylic acid (two COOH on benzene)

20. Which compound is salicylic acid?
    a. CH₃CH(OH)COOH   b. HOOC-COOH   c. benzene-1,2-dicarboxylic acid (two COOH)   d. benzene with COOH and OH (ortho)

21. For which pure compound is hydrogen bonding not possible?
    a. CH₃C(=O)-O-CH₃   b. CH₃C(=O)-OH   c. CH₃C(=O)-NHCH₃   d. CH₃CH₂OH

22. Which compound has the highest boiling point?
    a. CH₃CH₂O-CH₂CH₃   b. CH₃C(=O)-O-CH₃   c. CH₃C(=O)CH₂CH₃   d. CH₃CH₂CH₂CH₂OH
    e. CH₃CH₂C(=O)-OH

23. Which compound has the lowest boiling point?
    a. HO-CH₂CH₂CH₂OH   b. CH₃CH₂C(=O)-OH   c. CH₃C(=O)-O-CH₃

24. The odors associated with the smaller members of this class of compounds are distinctly unpleasant.
    a. amides   b. carboxylic acids   c. esters

25. The aromas associated with this class of compounds are regarded, in general, as pleasant.
    a. amides   b. carboxylic acids   c. esters

26. If CH₃CH₂C(=O)-O-CH₂CH₃ is hydrolyzed in base, which set of products is formed?

    a. CH₃CH₂OH   b. CH₃CH₂O⁻Na⁺   c. CH₃CH₂OH   d. CH₃CH₂O⁻Na⁺
       CH₃CH₂COOH   CH₃CH₂COOH   CH₃CH₂COO⁻Na⁺   CH₃CH₂COO⁻Na⁺

27. Esters can be synthesized from carboxylic acids and alcohols in the presence of mineral acids. Which does not improve the yield of ester by this reaction?
    a. addition of a large excess of the alcohol
    b. addition of a large excess of the carboxylic acid
    c. addition of a large excess of water
28. Aspirin is
    a. diacetylmorphine   b. acetylsalicylic acid   c. acetaminophen
29. ASA contains
    a. aspirin   b. aspirin, caffeine, phenacetin   c. acetaminophen, caffeine, phenacetin
30. Consider these compounds: a. heroin   b. nicotinamide   c. N,N-diethyllysergamide
    A. Which compound is classified as a vitamin?   a   b   c
    B. Which compound is classified as a narcotic?   a   b   c
    C. Which compound is classified as an hallucinogen? a   b   c

Functional Groups - Identify the functional groups in each of the following compounds. There may be more than one functional group in a single compound. The same functional group may appear in more than one compound.

List of Functional Groups:
a. alcohol
b. aldehyde
c. amide
d. carboxylic acid
e. ester
f. ether
g. ketone
h. phenol

Compound | Functional Groups Present

1. CH$_3$CH(OH)-C(=O)-OCH$_3$ a b c d e f g h
2. CH$_3$CH$_2$OCH$_2$CH$_2$OH   a b c d e f g h
3. CH$_3$OCH$_2$C(=O)-OH   a b c d e f g h
4. CH$_3$C(=O)-O-CH$_2$CH$_2$OH   a b c d e f g h
5. CH$_3$C(=O)-O-CH$_2$C(=O)-H   a b c d e f g h
6. CH$_3$O-C(=O)-CH$_2$C(=O)-NH$_2$ a b c d e f g h
7. HO-CH$_2$CH$_2$C(=O)-OH   a b c d e f g h
8. (salicylic acid: phenol with OH and COOH on benzene) a b c d e f g h
9. CH$_3$CH$_2$O-C$_6$H$_4$-NHC(=O)CH$_3$ a b c d e f g h
10. (steroid structure with O-C(=O)-CH$_3$ group)   a b c d e f g h

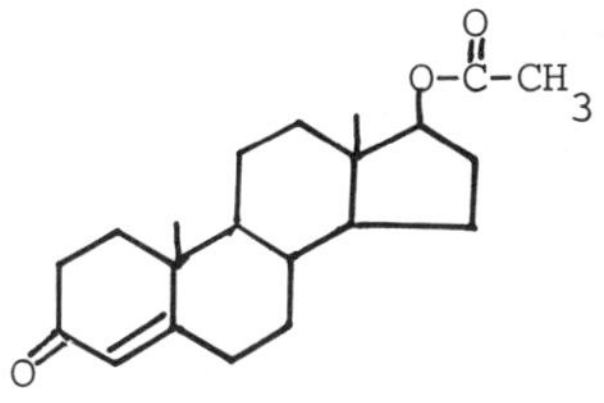

ANSWERS

Reactions:
1. 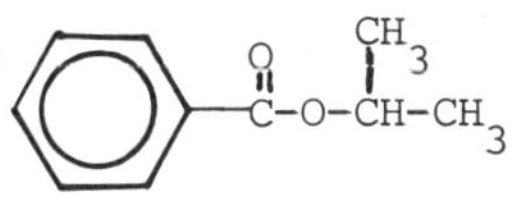 C$_6$H$_5$-C(=O)-O-CH(CH$_3$)-CH$_3$
2. CH$_3$-C(=O)-O-C$_6$H$_5$
3. (steroid with O-C(=O)-CH$_3$)
4. CH$_3$O-C(=O)CH$_2$CH$_2$C(=O)-O-CH$_3$
5. CH$_3$C(=O)-NH-C$_6$H$_5$ 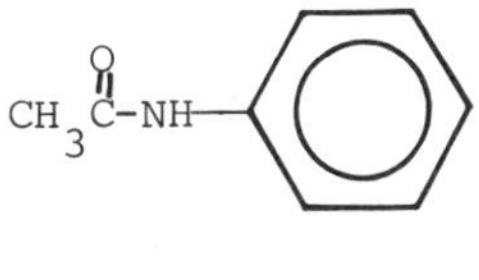
6. 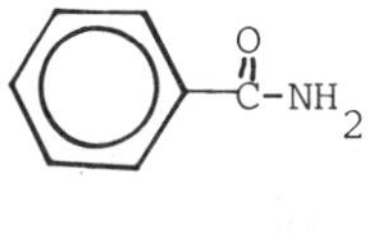 C$_6$H$_5$-C(=O)-NH$_2$

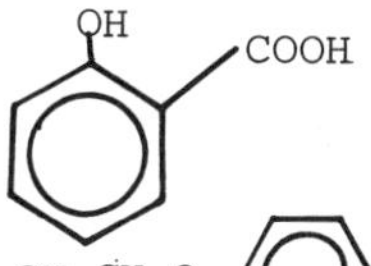

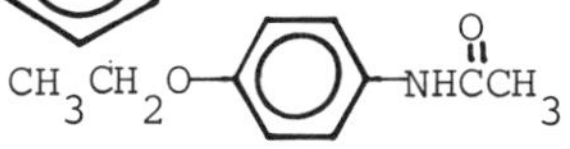

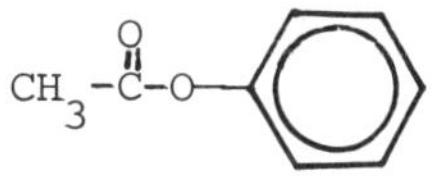

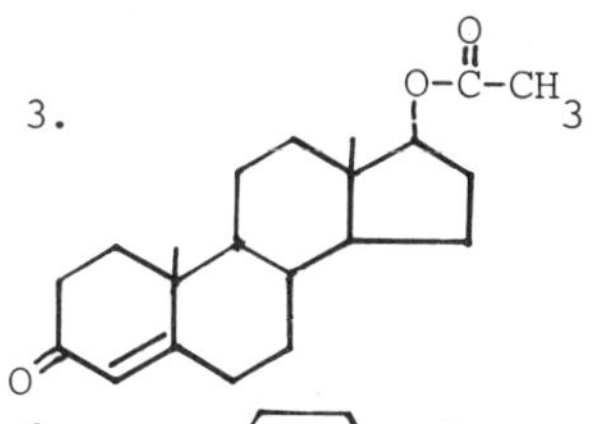

7. CH₃CHC(=O)-N(CH₃)-CH₃ with CH₃ branch

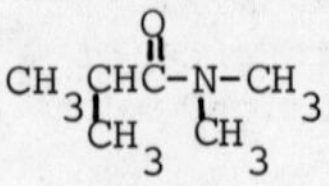

8. CH₃CH₂C(=O)-O⁻ Na⁺

9. benzene with two COO⁻Na⁺ groups (ortho)

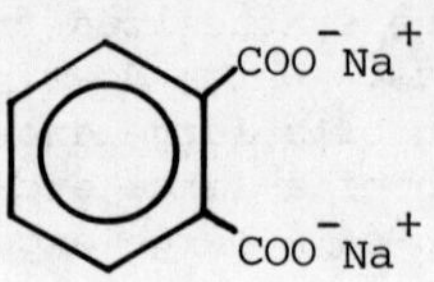

10. CH₃CH₂C(=O)-O⁻
    CH₃CH₂OH

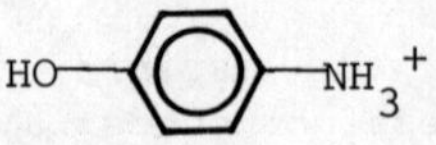

11. CH₃CH₂C(=O)-OH
    CH₃CH₂OH

12. HO-C₆H₄-NH₃⁺
    CH₃COOH

13. pyridine-3-C(=O)-O⁻ and NH₃

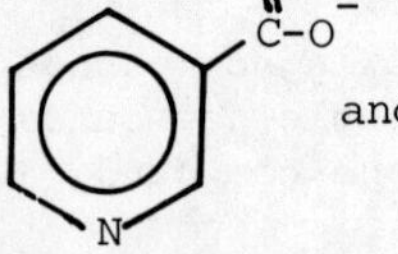

Self-Test:
1.	f	11.	c	21.	a
2.	a	12.	c	22.	f
3.	c	13.	c	23.	c
4.	d	14.	d	24.	b
5.	c	15.	c	25.	c
6.	d	16.	a	26.	c
7.	a	17.	b	27.	c
8.	a	18.	a	28.	b
9.	a	19.	b	29.	a
10.	a	20.	d	30.	A. b, B. a, C. c

Functional Groups:
1. a,e
2. a,f
3. d,f
4. a,e
5. b,e
6. c,e
7. a,d
8. d,h
9. c,f
10. e,g

Chapter 19 Amines and Derivatives

The questions at the end of chapter 19 review the chemistry and nomenclature introduced in the chapter. The few nomenclature rules were presented in section 19.2 and, since there are so few, we shall not both summarizing them here. However, since the general chemistry of the amines was scattered throughout several sections, we will offer a summary here.

General Chemistry of Amines

Synthesis of amines
    Chloride substitution: $R\text{-}Cl \xrightarrow{NH_3} R\text{-}NH_2$

    Reductive amination: $\underset{O}{R\text{-}\overset{\|}{C}\text{-}R} \xrightarrow{NH_3,\ H_2,\ Ni} R\text{-}\underset{\underset{NH_2}{|}}{CH}\text{-}R$

Reactions of amines
    Salt formation - All classes of amines (primary, secondary and tertiary) react.

$$R\text{-}NH_2 \xrightarrow{HCl} RNH_3^+\ Cl^-$$
$$R_2NH \xrightarrow{HCl} R_2NH_2^+\ Cl^-$$
$$R_3N \xrightarrow{HCl} R_3NH^+\ Cl^-$$

    Amide formation - Only primary and secondary amines react to give substituted amides.

        From acid chlorides: $RNH_2 + Cl\text{-}\overset{O}{\overset{\|}{C}}\text{-}R' \longrightarrow R\text{-}NH\text{-}\overset{O}{\overset{\|}{C}}\text{-}R'$

                             $R_2NH + Cl\text{-}\overset{O}{\overset{\|}{C}}\text{-}R' \longrightarrow R\text{-}\underset{\underset{R}{|}}{N}\text{-}\overset{O}{\overset{\|}{C}}\text{-}R'$

        From acid anhydrides: $RNH_2 + CH_3\overset{O}{\overset{\|}{C}}\text{-}O\text{-}\overset{O}{\overset{\|}{C}}CH_3 \longrightarrow R\text{-}NH\text{-}\overset{O}{\overset{\|}{C}}\text{-}CH_3$

                               $R_2NH + CH_3\overset{O}{\overset{\|}{C}}\text{-}O\text{-}\overset{O}{\overset{\|}{C}}CH_3 \longrightarrow R\text{-}\underset{\underset{R}{|}}{N}\text{-}\overset{O}{\overset{\|}{C}}\text{-}CH_3$

    Unsubstituted amides are commercially prepared by heating the ammonium salts of carboxylic acids.

$$R\text{-}\overset{O}{\overset{\|}{C}}\text{-}OH \xrightarrow{NH_3} R\text{-}\overset{O}{\overset{\|}{C}}\text{-}O^-\ NH_4^+ \xrightarrow{heat} R\text{-}\overset{O}{\overset{\|}{C}}\text{-}NH_2$$

Reaction with nitrous acid--Each class of amines reacts with nitrous acid in its own way.

    Primary amines yield nitrogen gas and a variety of organic products. It is the nitrogen gas which is measured in the Van Slyke method for the quantitative determination of primary amino groups.

    Secondary amines produce N-nitroso compounds: $R_2NH \xrightarrow{HO\text{-}N=O} R_2N\text{-}N=O$

    Tertiary amines may undergo a variety of reactions, but none (except salt formation) is important.

A major portion of the chapter was devoted to a consideration of physiologically important amines. The listing below offers a brief summary of this material.

Physiologically Active Compounds

Compound	Common Structural Feature	Some Effects

Hormones affecting mood
   epinephrine (adrenalin)
   norepinephrine
   serotonin
	see section 19.7	Balances involving these compounds produce states ranging from aggression to depression

Compound	Common Structural Feature	Some Effects

Amphetamines
  amphetamine
  methamphetamine
	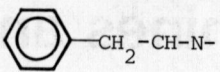 Ph-CH₂-CH-N-	stimulants; can cause insomnia, tremors, hallucinations, psychoses

Barbiturates
  barbital
  pentobarbital
  secobarbital
  phenobarbital
  amobarbital
  thiopental
	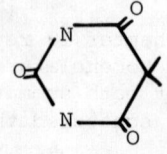 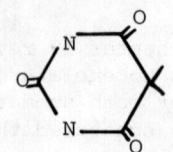 (barbiturate ring structure)	sedatives and soporifics; strongly addictive

Alkaloids
  Opium Alkaloids and Derivatives
    morphine
    codeine
    heroin
		narcotics; some are strongly addictive

coniine (from hemlock)	The alkaloids are nitrogen-containing heterocyclic compounds of varying degrees of complexity.	nausea, paralysis and death
caffeine (from coffee beans, tea leaves, cola nuts)		stimulant
nicotine (from tobacco)		stimulant
cocaine (from coca leaves)		powerful stimulant

Local Anesthetics
  benzocaine
  butesin
  tetracaine
  procaine
	 -O-C(=O)-Ph-N(H)-	localized insensitivity to pain

Tranquilizers
  Carbamates
    ethyl carbamate
    carisoprodol
    meprobamate
	-O-C(=O)-N-	calmatives and mild soporifics

Benzodiazepines
    diazepam
    chlordiazepoxide
	see section 19.12	

Promazines
    promazine
    chlorpromazine
    thioridazine
	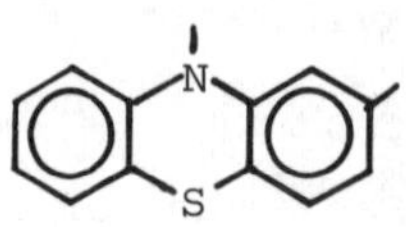 (phenothiazine ring)	

Self-Test

1. An example of a secondary amine is
   a. $C_2H_5NH_2$ b. $(C_2H_5)_2NH$ c. $(C_2H_5)_3N$ d. $(C_2H_5)_4N^+$

2. Which compound is a secondary amine?
   a.  4-methylaniline (NH_2 on benzene ring with CH_3 para)
   b. 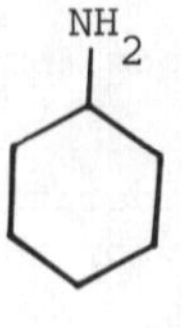 cyclohexylamine (NH_2 on cyclohexane)
   c. 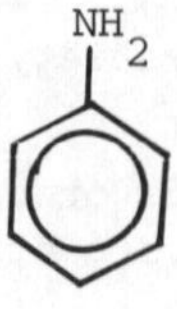 aniline (NH_2 on benzene)
   d. 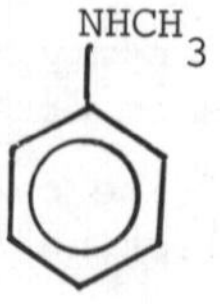 N-methylaniline ($NHCH_3$ on benzene)

3. Which compound contains a primary amine and a secondary alcohol group?

   a. $CH_2(OH)-CH(NH_2)-CH_2-CH_3$

   b. $CH_2(NHCH_3)-CH_2-CH(OH)-CH_3$

   c. $CH_3-CH(OH)-CH(NH_2)-CH_3$

4. 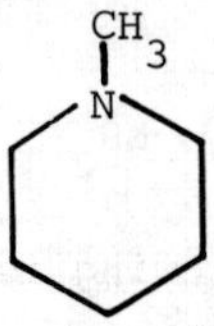 [N-methylpiperidine structure] is a(n):
   a. aniline   b. primary amine   c. secondary amine   d. tertiary amine

5. CH₃-CH(NH₂)-C(=O)-OH is a(n):
   a. amide   b. primary amine   c. secondary amine   d. tertiary amine

6. CH₃-NH-CH₂CH₃ is:
   a. 2-aminopropane   b. ethylmethylamine   c. isopropylamine   d. methyldimethylamine

7. CH₃-C(NH₂)(CH₃)-CH₂CH₃ is:
   a. isopentylamine   b. ethyldimethylamine   c. 2-aminoisopentane   d. 2-amino-2-methylbutane

8. 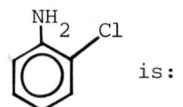 [o-chloroaniline structure: benzene with NH₂ and Cl ortho] is:
   a. o-chloroaniline   b. aniline hydrochloride   c. anilinium chloride

9. Aniline hydrobromide is
   a. 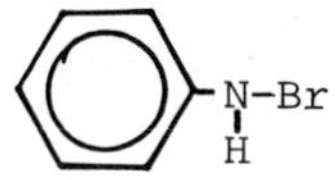 [C₆H₅-N(H)-Br]   b. 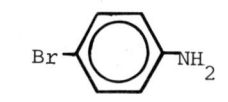 [Br-C₆H₄-NH₂ para]   c. 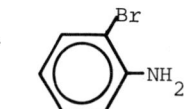 [Br-C₆H₄-NH₂ ortho]   d. 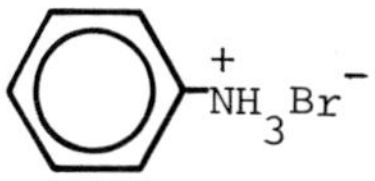 [C₆H₅-NH₃⁺ Br⁻]

10. Which is a quaternary ammonium ion?
    a. [CH₃-NH₃]⁺   b. [CH₃-NH₂-CH₃]⁺   c. [CH₃-NH(CH₃)-CH₃]⁺   d. [CH₃-N(CH₃)(CH₃)-CH₃]⁺

11. 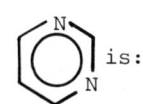 [pyrimidine ring with two N's] is:
    a. benzene   b. purine   c. pyridine   d. pyrimidine

12. Which compound would be expected to exist as a pair of mirror image isomers?
    a. 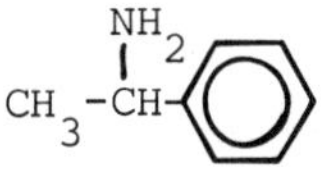 CH₃-CH(NH₂)-C₆H₅   b. 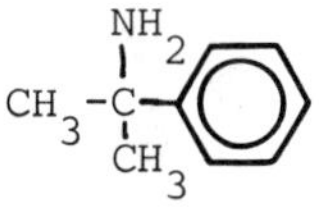 CH₃-C(NH₂)(CH₃)-C₆H₅   c. CH₃-C(=O)-C₆H₅   d.  CH₃-C₆H₄-NH₂

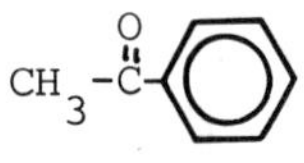

13. Alkaloids can best be generally classified with the
    a. alcohols   b. acids   c. amines   d. amides   e. esters

14. Which of the following compounds when pure is not associated through hydrogen bonding?
    a. CH₃CH₂OH   b. CH₃C(=O)OH   c. CH₃C(=O)NH₂   d. CH₃NH-CH₃   e. CH₃C(=O)NH-CH₃   f. CH₃N(CH₃)-CH₃

15. Which has the highest boiling point?
    a. CH₃CH₂CH₂OH   b. CH₃CH₂CH₂NH₂   c. CH₃NH-CH₂CH₃   d. CH₃N(CH₃)-CH₃

16. Which of the compounds shown in question 15 has the lowest boiling point?
    a   b   c   d

17. Some drugs which contain amino groups are converted to their salts to increase their
    a. acidity   b. basicity   c. solubility

18. Amines react with acids to form
    a. amino acids   b. esters   c. bases   d. salts

19. Which compound does not exhibit the properties of a base?

    a. $CH_3CH_3CH_2NH_2$ b. CH_3CHCH_3 with NH_2 c. $CH_3-\underset{O}{\overset{\|}{C}}-NH_2$ d. $CH_3-\underset{CH_3}{\overset{|}{N}}-CH_3$ e. 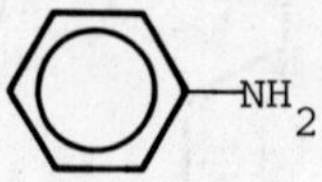 $C_6H_5-NH_2$

20. If a carboxylic acid and an amine react to form a salt, which of the following is the product?

    a. $RCOO^-\ RNH_3^+$ b. $RCOO^+\ RNH_3^-$ c. $RCOOH^+\ RNH_2^-$ d. $RCOOH^+\ RNH_2^-$

21. A tertiary amine does not react with
    a. HCl   b. $CH_3\underset{O}{\overset{\|}{C}}-O-\underset{O}{\overset{\|}{C}}CH_3$ c. HO-N=O

22. The product of the reaction between isopropyl chloride and ammonia is

    a. $CH_3-\underset{NH_2}{\overset{Cl}{\underset{|}{\overset{|}{C}}}}-CH_3$ b. $CH_3-\underset{NH_2}{\overset{|}{CH}}-CH_3$ c. $CH_3-\overset{O}{\overset{\|}{C}}-NH_2$

23. The reaction between acetic anhydride and aniline produces

    a. 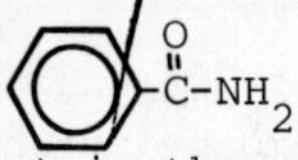 $C_6H_5-\overset{O}{\overset{\|}{C}}-NH_2$ b. 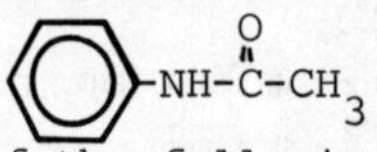 $C_6H_5-NH-\overset{O}{\overset{\|}{C}}-CH_3$ c. $CH_3\overset{O}{\overset{\|}{C}}-NH_2$ d. $CH_3\overset{O}{\overset{\|}{C}}-O^-\ NH_4^+$

24. What is the product of the following reaction: $CH_3\overset{O}{\overset{\|}{C}}-CH_3 \xrightarrow{H_2,\ NH_3,\ Ni}$?

    a. $CH_3-\overset{O}{\overset{\|}{C}}-NH_2$ b. $CH_3-\overset{O}{\overset{\|}{C}}-NH-CH_3$ c. $CH_3CH_2-\overset{O}{\overset{\|}{C}}-NH_2$ d. $CH_3-\underset{NH_2}{\overset{|}{CH}}-CH_3$

25. Which compound would not react with an acid chloride to form an amide?

    a. NH_3 b. CH_3NH_2 c. $CH_3-NH-CH_3$ d. $CH_3-\underset{CH_3}{\overset{|}{N}}-CH_3$ e. 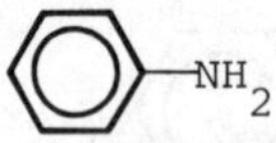 $C_6H_5-NH_2$

26. Which reaction is called reductive amination?

    a. $CH_3CH_2Cl \xrightarrow{NH_3} CH_3CH_2NH_2$

    b. $CH_3\overset{O}{\overset{\|}{C}}-CH_3 \xrightarrow{H_2,\ NH_3,\ Ni} CH_3\underset{NH_2}{\overset{|}{CH}}-CH_3$

    c. $CH_3-NH-CH_3 \xrightarrow{HNO_2} CH_3-\underset{CH_3}{\overset{|}{N}}-N=O$

27. Which reagent is used to detect the presence of amino acids through a color reaction?
    a. ninhydrin   b. pyrimidine   c. epinephrine

28. In the Van Slyke method, primary amino groups are detected by
    a. the development of a blue color
    b. the change of litmus paper from red to blue
    c. the evolution of nitrogen gas

29. Members of which class of compounds were not identified as carcinogens?
    a. N-nitroso compounds   b. aromatic amines   c. amino acids

30. Which compound is commonly referred to as adrenalin?
    a. serotonin   b. epinephrine   c. norepinephrine

31. The molecular theory of mental illness postulates an imbalance between two biochemicals. An excess of which compound produces mental depression?
    a. serotonin   b. epinephrine   c. norepinephrine

32. Which compound is classified as a "downer"?
    a. amphetamine   b. barbital   c. cocaine

33. The amphetamines are structurally related to
    a. the natural stimulants epinephrine and norepinephrine
    b. the alkaloids, cocaine and caffeine
    c. the tranquilizers Valium and Librium

34. Which molecular group is common to all barbiturates?

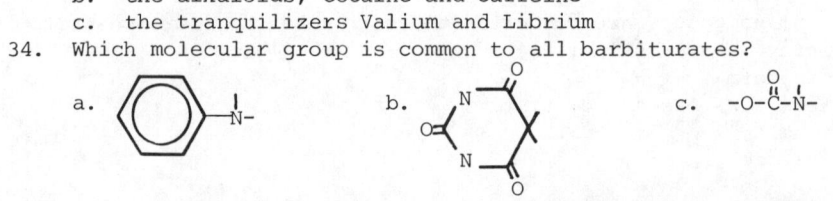

    a. C_6H_5-N- b. (barbiturate ring structure with N, N, and three C=O groups)   c. $-O-\overset{O}{\overset{\|}{C}}-N-$

35. Reserpine, lithium carbonate and the promazines are all used
    a. in the treatment of mental illnesses
    b. as local anesthetics
    c. as general anesthetics

ANSWERS:

1.	b	16.	d	31.	a
2.	d	17.	c	32.	b
3.	c	18.	d	33.	a
4.	d	19.	c	34.	b
5.	b	20.	a	35.	a
6.	b	21.	b		
7.	d	22.	b		
8.	a	23.	b		
9.	d	24.	d		
10.	d	25.	d		
11.	d	26.	b		
12.	a	27.	a		
13.	c	28.	c		
14.	f	29.	c		
15.	a	30.	b		

Chapter 20 Compounds of Sulfur and Phosphorus

First we'll call your attention to table 20.2 in the text. This table neatly summarizes the structural variations and the nomenclature of the sulfur compounds encountered in the chapter. The organic compounds incorporating phosphorus are not so easily tabulated. Two general classes, the phosphines and the phosphate esters, were presented in the chapter. However, many of the more interesting phosphorus-containing organic compounds do not fall under these simple classifications. In some ways they appear to be molecules which might have started as phosphate esters, but which have been modified so much that they no longer fit that category.

Organophosphorus Compounds

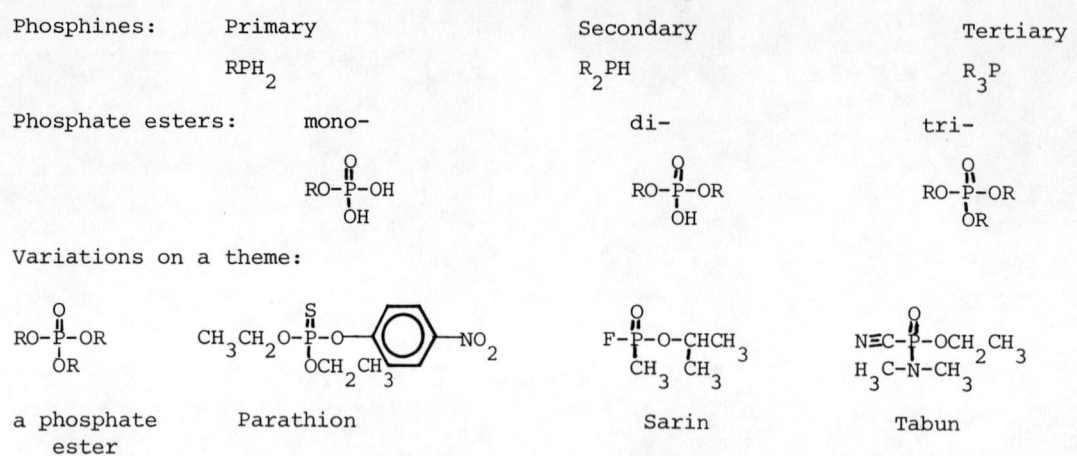

Phosphines: Primary   Secondary   Tertiary

$\qquad\qquad\qquad$ RPH_2 $\qquad\qquad\qquad\qquad$ R_2PH $\qquad\qquad\qquad\qquad$ R_3P

Phosphate esters: mono-   di-   tri-

$$RO-\overset{\overset{O}{\|}}{P}-OH \qquad RO-\overset{\overset{O}{\|}}{P}-OR \qquad RO-\overset{\overset{O}{\|}}{P}-OR$$
$$\,\,\,\,OH \qquad\qquad\quad\,\,\, OH \qquad\qquad\quad\,\, OR$$

Variations on a theme:

$$RO-\overset{\overset{O}{\|}}{P}-OR \qquad CH_3CH_2O-\overset{\overset{S}{\|}}{P}-O-\!\!\bigcirc\!\!-NO_2 \qquad F-\overset{\overset{O}{\|}}{P}-O-CHCH_3 \qquad N\!\!\equiv\!\!C-\overset{\overset{O}{\|}}{P}-OCH_2CH_3$$
$$\,\,\,OR \qquad\qquad\qquad OCH_2CH_3 \qquad\qquad\quad CH_3\,\,CH_3 \qquad\qquad H_3C-N-CH_3$$

a phosphate $\qquad$ Parathion $\qquad\qquad\qquad$ Sarin $\qquad\qquad\qquad$ Tabun
ester

We examined some of the chemistry of organosulfur compounds. These interrelated reactions are diagrammed below. Remember the [O] means oxidation and [H] means reduction. When a choice of reagents is possible, the selection is presented in parentheses.

Reactions of Organosulfur Compounds

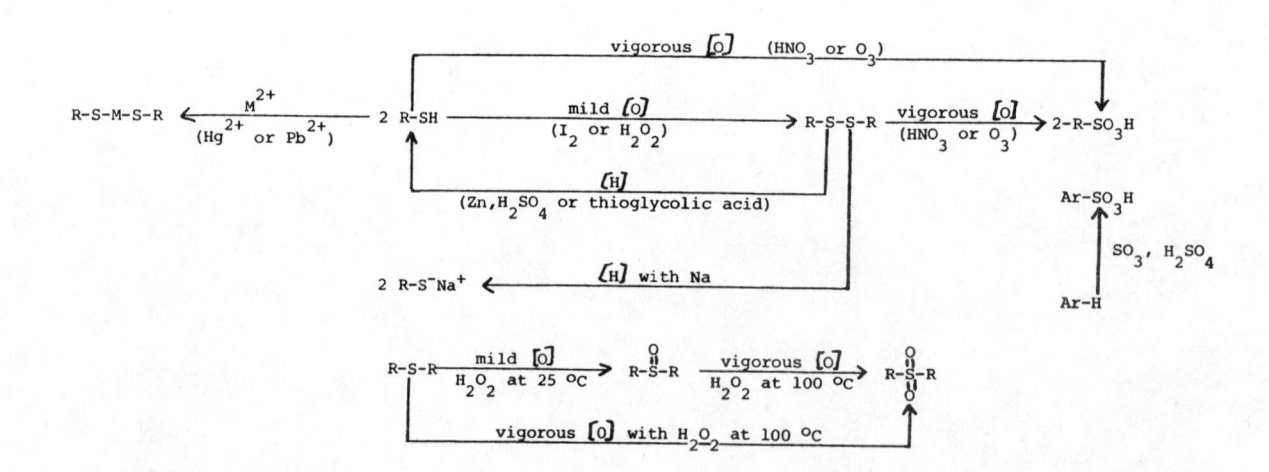

$$R\text{-}S\text{-}M\text{-}S\text{-}R \xleftarrow{\underset{(Hg^{2+}\text{ or }Pb^{2+})}{M^{2+}}} 2\,R\text{-}SH \xrightarrow[\text{(I_2 or H_2O_2)}]{\text{mild [O]}} R\text{-}S\text{-}S\text{-}R \xrightarrow[\text{(HNO_3 or O_3)}]{\text{vigorous [O]}} 2\text{-}R\text{-}SO_3H$$

vigorous [O] (HNO_3 or O_3) (overall arrow from 2 R-SH to 2-R-SO_3H)

[H] (Zn, H_2SO_4 or thioglycolic acid)

$2\,R\text{-}S^-Na^+ \xleftarrow{\text{[H] with Na}}$

$Ar\text{-}SO_3H \xleftarrow{SO_3, H_2SO_4} Ar\text{-}H$

$$R\text{-}S\text{-}R \xrightarrow[H_2O_2 \text{ at } 25\,°C]{\text{mild [O]}} R\text{-}\overset{\overset{O}{\|}}{S}\text{-}R \xrightarrow[H_2O_2 \text{ at } 100\,°C]{\text{vigorous [O]}} R\text{-}\overset{\overset{O}{\|}}{\underset{\underset{O}{\|}}{S}}\text{-}R$$

vigorous [O] with H_2O_2 at 100 °C

Problems 1 and 2 at the end of the chapter in the text thoroughly test your knowledge of nomenclature. Problem 3 covers the chemistry of the sulfur compounds. The self-test reviews this material and the remaining sections of the chapter.

Self-Test

1. sec-Butyl mercaptan is
   a. 2-butanesulfide b. 2-butylsulfide c. butyl thiol d. 2-butanethiol
2. $CH_3CH_2S-S-CH_2CH_2SH$ does not contain a
   a. disulfide group b. sulfide group c. sulfhydryl group
3. $CH_3\overset{CH_3}{\underset{|}{CH}}-S-\overset{CH_3}{\underset{|}{CH}}-CH_3$ is
   a. isopropyl disulfide b. diisopropyl sulfide c. isopropyl mercaptan
4. The sulfhydryl group is
   a. $-SO_3H$ b. $-S-OH$ c. $-SH$ d. $-S-S-$ e. $-S-$
5. 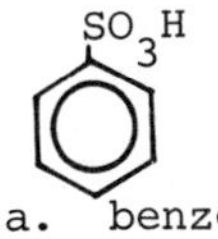 [benzene ring with SO_3H] is
   a. benzenesulfonic acid b. benzenethiol c. benzenesulfoxide d. benzenesulfone
6. Which compound is trimethyl phosphate?
   a. $CH_3O-\overset{O}{\underset{|}{P}}(OH)-OH$ b. $CH_3O-\overset{O}{\underset{|}{P}}(OCH_3)-OCH_3$ c. $CH_3\overset{}{\underset{CH_3}{P}}CH_3$ d. $CH_3-\overset{O}{\underset{CH_3}{P}}-CH_3$ e. $CH_3O-\overset{O}{\underset{OH}{P}}-O-\overset{O}{\underset{OH}{P}}-O-\overset{O}{\underset{OH}{P}}-OH$
7. Which compound listed in question 6 is an ester of triphosphoric acid?
   a b c d e
8. Which compound listed in question 6 is a phosphine?
   a b c d e
9. Dimethyl sulfone is
   a. $CH_3-\overset{O}{\underset{O}{S}}-CH_3$ b. CH_3-S-CH_3 c. $CH_3-\overset{O}{S}-CH_3$ d. $CH_3-\overset{O}{\underset{OH}{S}}-CH_3$ e. $CH_3-S-S-CH_3$
10. Which compound listed in question 9 is a sulfoxide?
    a b c d e
11. What is the product of the reaction: $2\ CH_3-SH\ \xrightarrow{mild[O]}$?
    a. CH_3SCH_3 b. CH_3S-SCH_3 c. $CH_3\overset{O}{S}CH_3$ d. $2\ CH_3SO_3H$
12. Which reagent will bring about the following reaction: $CH_3S-S-CH_3 \longrightarrow 2\ CH_3SH$?
    a. Zn, H_2SO_4 b. H_2O_2 c. SO_3, H_2SO_4 d. I_2
13. Which reaction does not produce benzenesulfonic acid?
    a. 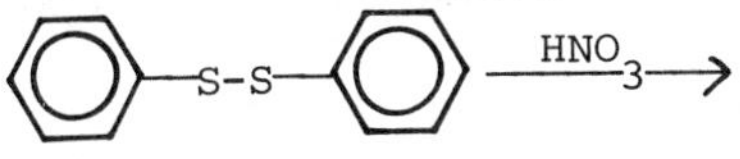 Ph-S-S-Ph $\xrightarrow{HNO_3}$
    b. 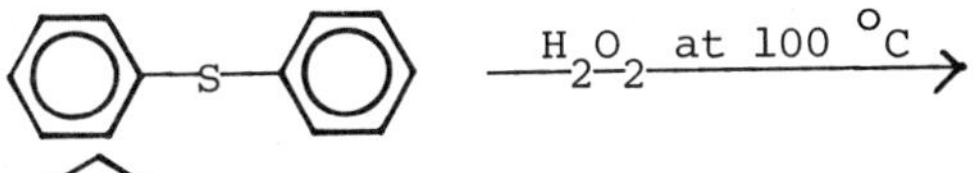 Ph-S-Ph $\xrightarrow{H_2O_2\ at\ 100\ °C}$
    c. Ph $\xrightarrow{SO_3,\ H_2SO_4}$

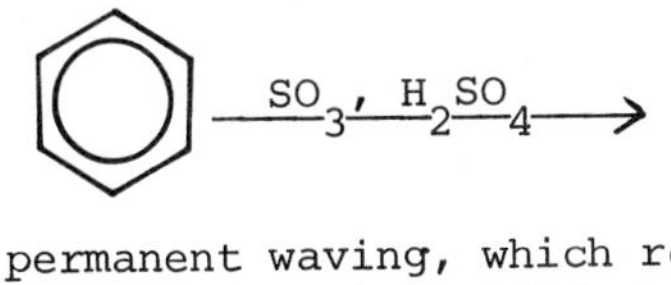

14. In permanent waving, which reaction is carried out first?
    a. oxidation of disulfide bonds
    b. reduction of disulfide bonds
    c. oxidation of sulfhydryl groups
    d. reduction of sulfhydryl groups
15. Mercury ions react with
    a. sulfhydryl groups b. sulfide bonds c. sulfone groups
16. Treatment of diphenyl disulfide with sodium metal yields
    a. 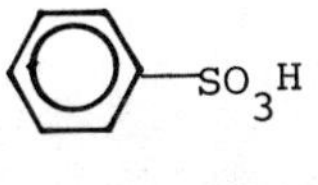 Ph-SO_3H b. 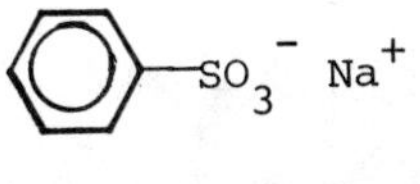 Ph-$SO_3^-Na^+$ c. 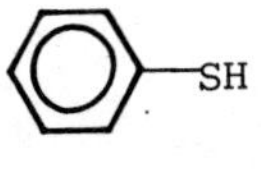 Ph-SH d. 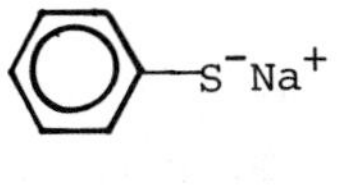 Ph-S^-Na^+

17. Which is the strongest acid?

    a. C₆H₅—SH   b. C₆H₅—SO₃H   c. C₆H₅—COOH   d. C₆H₅—OH

18. The odors of the thiols (are)
    a. pleasant   b. unpleasant   c. depend on the molecular weights of the compounds
19. Although toxic, the low molecular weight thiols are not as dangerous as some poison because
    a. they are not volatile
    b. they can be detected at very low concentrations by their odors
    c. they are deactivated by mercury ions in the blood stream
20. BAL is an effective antidote for mercury poisoning because
    a. BAL chelates mercury ions
    b. BAL reduces the mercury ions to metallic mercury
    c. BAL reacts with the enzymes that would otherwise react with the mercury
21. Which is not true of dimethyl sulfoxide?
    a. It both dissolves in water and dissolves organic compounds.
    b. It acts as an analgesic and an antiinflammatory agent.
    c. It is used as the reducing agent of permanent waving solutions.
22. The LAS detergents do not incorporate
    a. the salt of a sulfonic acid
    b. highly branched, biodegradable aliphatic side chains
    c. a benzene ring carrying two substituents
23. The sulfa drugs are
    a. sulfides   b. disulfides   c. sulfonic acids   d. sulfonamides
24. Which compounds are not toxic to humans?
    a. alkyl phosphines
    b. the phosphorylating agents ATP and ADP
    c. the nerve gases, Agents GB and VX
25. Phosphorus-containing compounds serve as
    a. pesticides   b. nerve gases
    c. intermediates in carbohydrate   d. a and b
       metabolism   e. a, b and c
26. Which compound is responsible for transmitting an impulse across a nerve synapse?
    a. choline   b. acetylcholine   c. cholinesterase
27. Which of the compounds listed in question 26 is the enzyme responsible for resetting the receptor to "off" after transmission of a signal across the synapse?
    a   b   c
28. The most toxic substance known, botulinus, acts by
    a. preventing the synthesis of acetylcholine
    b. blocking the receptor site to acetylcholine
    c. preventing the breakdown of acetylcholine
29. Which of the items listed in question 28 is the mode of action of anticholinesterases?
    a   b   c
30. Which of the items listed in question 28 is the mode of action of some local anesthetics?
    a   b   c

ANSWERS:

1. d	11. b	21. c
2. b	12. a	22. b
3. b	13. b	23. d
4. c	14. b	24. b
5. a	15. a	25. e
6. b	16. d	26. b
7. e	17. b	27. c
8. c	18. c	28. a
9. a	19. b	29. c
10. c	20. a	30. b

Chapter 21 Polymers

Condensation polymerization involves chemistry we have encountered many times, particularly in the reactions of alcohols, amines and carboxylic acids. Thus, polyester formation differs from simple esterification only in the number of **functional groups** per molecule which react. In simple esterification, an alcohol with one hydroxyl group reacts with an acid with one carboxyl group:

$$R-\underset{O}{\overset{\|}{C}}-OH + HO-R' \longrightarrow R-\underset{O}{\overset{\|}{C}}-O-R' + H_2O$$

In polymerization, the only difference is that each molecule must have at least 2 functional groups. In polyester formation, for example, one molecule may have 2 hydroxyl groups and the other 2 carboxyl groups or every molecule may have one hydroxyl and one carboxyl group:

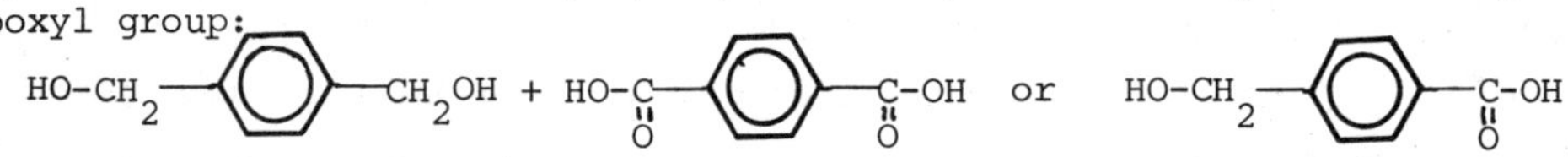

$$HO-CH_2-\bigcirc-CH_2OH + HO-\underset{O}{\overset{\|}{C}}-\bigcirc-\underset{O}{\overset{\|}{C}}-OH \quad \text{or} \quad HO-CH_2-\bigcirc-\underset{O}{\overset{\|}{C}}-OH$$

The reaction of both functional groups on each molecule ties hundreds or thousands of these molecules together to form the polymer chain.

Addition polymerization involves chemistry similar to but not precisely the same as the chemistry we previously discussed. We have said that the characteristic reaction of alkenes is the addition reaction. We have usually added to the double bond small inorganic molecules such as Br_2 or H_2 or H_2O. Now we are proposing the addition of one alkene molecule to another. Here's more detail on how that works.

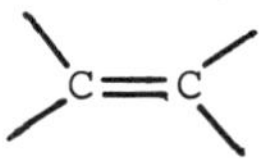

>C=C<

This is the alkene molecule. All sorts of groups can be attached to the doubly bonded carbons (just the situation one needs to make a variety of polymers). For the moment, however, we're going to ignore everything except the two carbons in the double bond.

C⋯C

Also, we're going to redraw the double bond with its 2 pairs of shared electrons so that one pair of electrons is represented by the usual line, but the second pair is drawn explicitly (as electron dots). Remember that there are still 4 other bonds to the 2 carbons in the double bond; we're just not drawing them in at the moment.

C⋯C C⋯C C⋯C C⋯C C⋯C

Here is a group of monomer molecules.

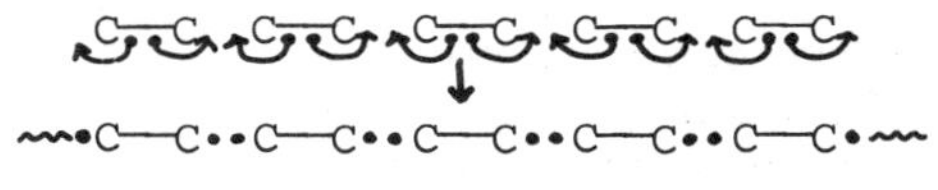

As they notice one another they join hands, which in molecule-language means they shift the electrons in the double bond to both sides and begin sharing electrons with the adjacent molecules.

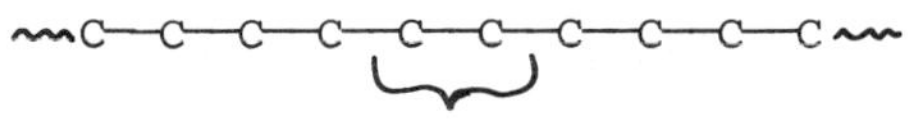

~•C—C••C—C••C—C••C—C••C—C•~

If we replace the shared pairs of electron dots with bond lines, we get the usual picture of the polymer...

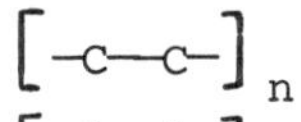

~C—C—C—C—C—C—C—C—C—C~

...which can be translated to the abbreviated form...

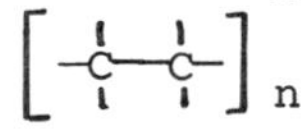

$$[-C-C-]_n$$

...which should be drawn with the four additional bonds we eliminated earlier.

$$\left[-\overset{|}{\underset{|}{C}}-\overset{|}{\underset{|}{C}}-\right]_n$$

84

Elastomers are usually formed from molecules which have 2 double bonds to start with. The monomer for the synthetic rubber, poly-chloroprene, is 2-chloro-1,3-butadiene:

This structural formula shows the proper spacial arrangement of bonds.

But we can draw a straight line formula which still conveys the proper attachment of atoms.

We can simplify this even more by not drawing in all of the bonded hydrogens. These atoms are still there, but now they're "understood."

One last change before we begin drawing the polymerization reaction. Again, let's replace one of the bond lines of each double bond with a pair of electron dots.

In polymerization, a group of these molecules get together.

The end carbons of adjacent molecules grab hands.

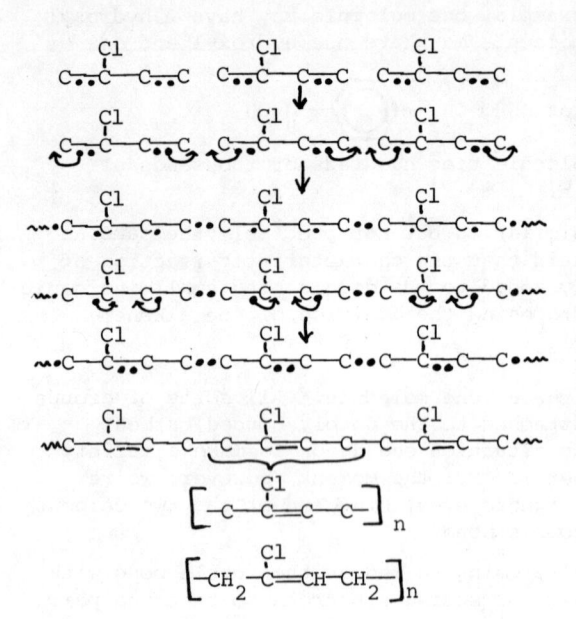

That process, unfortunately, leaves the two middle carbons of each unit with unshared electrons. They do not like this, but immediately recognize that the problem can be solved by sharing with each other.

Now let's replace the electron dots with bond lines...

...and translate to the abbreviated formula...

...and add all the missing hydrogens back.

That's it -- the chemistry of polymerization.

The terminology introduced in the chapter is reviewed in question 1 at the end of the chapter. Be sure to work through this problem before attempting the self-test.

Self-Test

1. The small-molecule starting materials from which macromolecules can be constructed are called
    a. monomers   b. polymers   c. segmers
2. In which method of polymerization are nonpolymeric byproducts formed?
    a. addition polymerization   b. condensation polymerization
3. Which compound would not serve as a monomer in addition polymerization?
    a. $CH_2=CH-COOH$ b. $CH_2=CH-CH_2OH$ c. $HO-CH_2-CH_2-COOH$
4. Would this reaction be classified as addition or condensation polymerization:

$$n\ CH_2\overset{O}{-}CH_2 \longrightarrow [-CH_2-CH_2-O-]_n$$

    a. addition   b. condensation
5. The polymer formed from $CH_2=\overset{Cl}{\underset{}{C}}-CH_3$ is

   a. $[-CH_2=\overset{CH_2Cl}{\underset{}{C}}-]_n$ b. $[-CH_2=\overset{Cl}{\underset{}{C}}-CH_3-]_n$ c. $[-CH_2-\overset{Cl}{\underset{}{C}}-CH_3]_n$ d. $[-CH_2-\overset{CH_2Cl}{\underset{}{CH}}-]_n$

   e. $[-CH_2-\overset{Cl}{\underset{CH_3}{C}}-]_n$

6. If the monomer is $CH_3-CH=CH-Cl$, the polymer is

   a. $\left[-CH_3-CH=CH-\right]_n$ with Cl b. $\left[-CH_3CH_2-CH-\right]_n$ with Cl c. $\left[-CH_3-CH=CH-Cl-\right]_n$ d. $\left[-CH-CH-\right]_n$ with H_3C and Cl

7. Which natural polymer serves as the starting material for rayon and acetate fibers?
   a. cellulose   b. silk   c. wool
8. Which product made by the chemical manipulation of cellulose exists as a derivative of cellulose instead of regenerated cellulose?
   a. cellulose acetate   b. rayon   c. cellophane
9. Which is the more highly ordered arrangement of matter?
   a. high density polyethylene   b. low density polyethylene
10. In chemical usage, a plastic is a substance which
    a. will return to its original shape after being stretched
    b. softens on heating and can be molded under pressure
    c. hardens under the influence of heat and pressure
11. Which material would best maintain its shape when heated?
    a. low density polyethylene
    b. a thermosetting polymer
    c. a cross-linked polymer
12. Which type of forces operate between the long polymer chains of polyethylene?
    a. hydrogen bonding
    b. dipole interactions
    c. dispersion forces
    d. ionic bonding
13. Which does not apply to the polymer Bakelite?
    a. It is a copolymer.
    b. It is a condensation polymer.
    c. It is a cross-linked polymer.
    d. It is an elastomer.
    e. All of the above apply to Bakelite.
14. Natural rubber is
    a. polyethylene   b. Neoprene   c. polyisoprene   d. poly(vinyl chloride)
15. Which is not true of the vulcanization process?
    a. It results in sulfur bridges between polymer chains.
    b. It increases the hardness of natural rubber.
    c. It improves the elasticity of natural rubber.
    d. All of the above apply to the vulcanization process.
16. Which natural fiber is not linked by the same functional group as nylon?
    a. cotton   b. silk   c. wool
17. What is the functional group linking the segmers of nylon?
    a. amide   b. disulfide   c. ester
18. All natural and synthetic fibers
    a. are condensation polymers
    b. are copolymers
    c. exhibit appreciable tensile strength
    d. All of the above are true.
19. Dipolar interactions between polymer chains are not found in
    a. polyamides   b. polyesters   c. polyethylenes
20. One approach to the development of flame-retardant fabrics is the use of polymer fibers which incorporate
    a. sulfur crosslinks
    b. chlorine and bromine atoms
    c. benzene rings
    d. plasticizers
21. Plasticizers are
    a. polymers which show elastic properties
    b. polymers which soften on heating
    c. molecules which confer pliability on otherwise brittle polymers
22. Plasticizers
    a. are part of the polymer chain
    b. form crosslinks from one polymer chain to another
    c. are inserted in spaces between polymer chains

23. Which compounds have not been used as plasticizers?
    a. esters of phthalic acid
    b. polychlorinated biphenyls
    c. vinyl chlorides
    d. All of the above have been used as plasticizers.
24. Synthetic polymers used for biomedical purposes are ordinarily coated with heparin to
    a. make the otherwise rigid polymers more plastic
    b. prevent the rejection of the artificial parts by the natural defenses of the body
    c. prevent blood from coagulating on contact with the synthetic material
    d. All of the above are true.
25. Disposal of synthetic "plastics" is complicated by
    a. the huge volume of these materials which accumulates as trash
    b. the non-biodegradable nature of most synthetic
    c. the toxicity of some gases produced when the materials are burned
    d. All of the above are valid.

ANSWERS

1.	a	11.	a	21.	c
2.	b	12.	c	22.	c
3.	c	13.	d	23.	c
4.	a*	14.	c	24.	c
5.	e	15.	d	25.	d
6.	d	16.	a		
7.	a	17.	a		
8.	a	18.	c		
9.	a	19.	c		
10.	b	20.	b		

---

*Note that it is not necessary for a monomer to contain a double bond for addition polymerization to occur. (It is, however, usual.) The definition of addition polymerization simply requires that all atoms of the monomer be incorporated in the polymer.

Chapter 22 Carbohydrates

In chapter 22 the terminology and molecular architecture associated with carbohydrates were presented. Our study of carbohydrates will continue in chapter 30 when we trace the critical biochemical role of these compounds. To do that, it is essential that you familiarize yourself now with carbohydrate structures. You should aim at being able to read carbohydrate structures as well as you now read the structures of simpler compounds. For example, when we say "the hydroxyl group of ethanol", with only the slightest pause you should think "CH_3-CH_2-OH." After studying the material in this chapter, you should also be able to look at a carbohydrate structure and, with a pause only a bit longer than before, locate the acetal linkage or determine whether the compound is glucose or fructose or mannose or whether or not it is a reducing sugar.

After you read the chapter, go through all of the questions at the end of the chapter. These questions have been designed to review in a step by step manner the names and structures of the carbohydrates. Only after you've completed that review should you attempt the self-test.

Before we present the self-test, let's establish a way of drawing saccharides in abbreviated form. For the open-chain form, we'll adopt this shorthand:

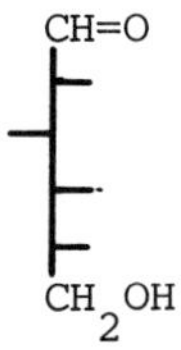

$$\begin{array}{c} CH=O \\ \vdash \\ \dashv \\ \vdash \\ CH_2OH \end{array} \quad \text{is the same as} \quad \begin{array}{c} H-C=O \\ H-C-OH \\ HO-C-H \\ H-C-OH \\ H-C-OH \\ CH_2OH \end{array}$$

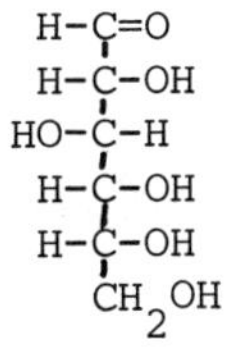

The branches on the stem of the abbreviated formula give the orientation of the secondary hydroxyl groups. The same simplification can be used for ring forms:

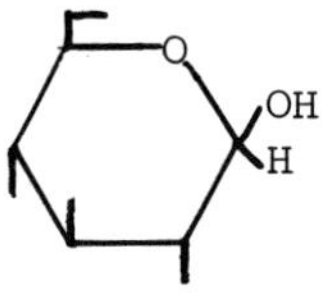

(ring structure) is the same as (ring structure with CH₂OH, OH, HO, H groups)

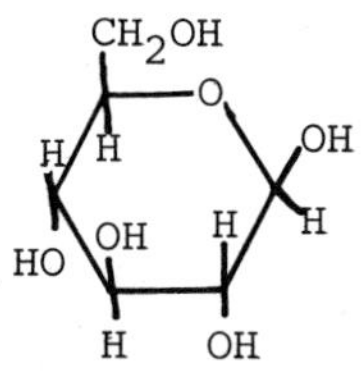

We'll use the simplified structures in the self-test. Be sure you're able to translate these forms to the more familiar ones before you proceed.

<u>Self-Test</u>

Refer to the following structures in answering questions 1 through 6.

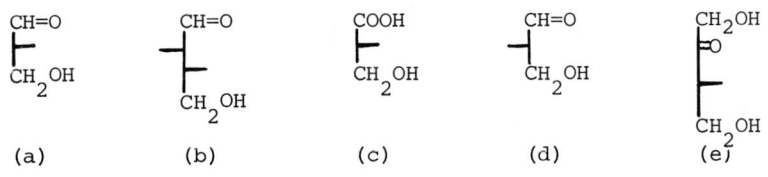

(a) CH=O / ⊢ / CH_2OH

(b) CH=O / ⊢ / ⊣ / CH_2OH

(c) COOH / ⊢ / CH_2OH

(d) CH=O / ⊣ / CH_2OH

(e) CH_2OH / ⊢=O / ⊢ / CH_2OH

1. Which of the compounds is the mirror image isomer of (a)?   a   b   c   d   e
2. Which is not a member of the D-family of sugars?   a   b   c   d   e
3. Which is D-glyceraldehyde?   a   b   c   d   e
4. Which is the product of oxidation of D-glyceraldehyde?   a   b   c   d   e
5. Which is an aldotetrose?   a   b   c   d   e
6. Which is a ketose?   a   b   c   d   e

Refer to the following structures in answering questions 7 through 14.

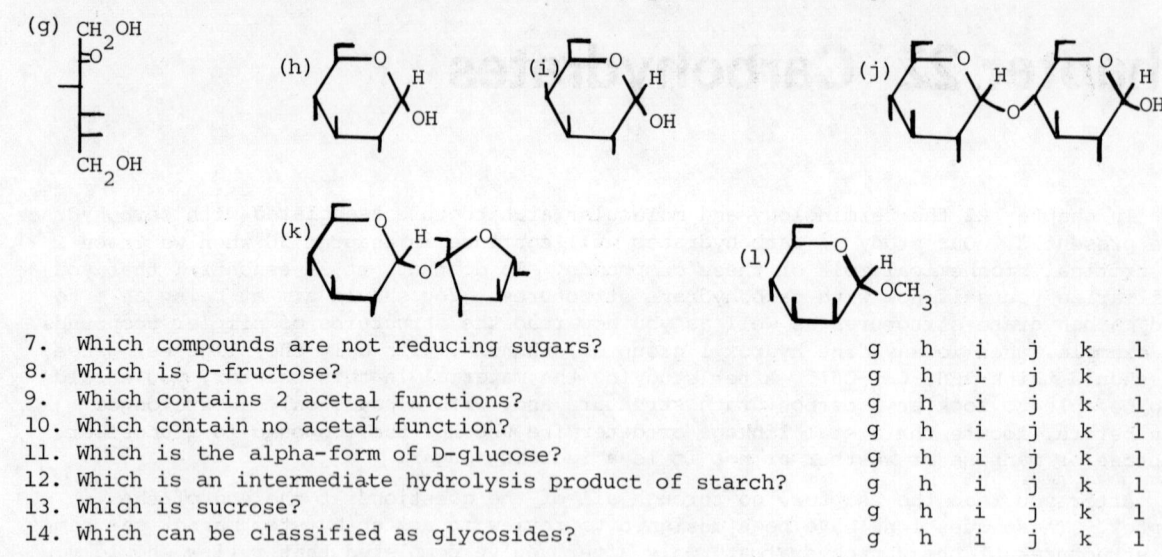

(g) CH₂OH–CO–(...)–CH₂OH   (h)   (i)   (j)   (k)   (l) OCH₃

7. Which compounds are not reducing sugars?   g h i j k l
8. Which is D-fructose?   g h i j k l
9. Which contains 2 acetal functions?   g h i j k l
10. Which contain no acetal function?   g h i j k l
11. Which is the alpha-form of D-glucose?   g h i j k l
12. Which is an intermediate hydrolysis product of starch? g h i j k l
13. Which is sucrose?   g h i j k l
14. Which can be classified as glycosides?   g h i j k l

Refer to the following structure in answering questions 15 through 22.

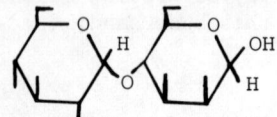

15. The monosaccharide unit on the right is
    a. fructose b. galactose c. glucose d. lactose e. maltose f. mannose
    g. sucrose   h. none of these
16. The monosaccharide unit on the left is
    a. fructose b. galactose c. glucose d. lactose e. maltose f. mannose
    g. sucrose   h. none of these
17. The disaccharide is
    a. fructose b. galactose c. glucose d. lactose e. maltose f. mannose
    g. sucrose   h. none of these
18. The hemiacetal group is: a. alpha   b. beta
19. The acetal group is: a. alpha   b. beta
20. The rings are attached through positions: a. 1,4 b. 1,6 c. ortho d. meta e. para
21. The compound is a: a. reducing sugar   b. nonreducing sugar
22. On hydrolysis, the compound yields
    a. aldohexoses   b. aldopentoses   c. ketohexoses   d. ketopentoses
23. Which of the following yields a product other than glucose upon complete hydrolysis?
    a. amylopectin   b. cellulose   c. lactose   d. maltose   e. starch
24. Common table sugar is more formally described as
    a. glucose   b. lactose   c. maltose   d. sucrose
25. Hydrolysis of which disaccharide gives glucose and fructose as products?
    a. cellulose   b. galactose   c. lactose   d. maltose   e. sucrose
26. Blood sugar is the same as
    a. fructose   b. galactose   c. glucose   d. glycogen   e. lactose   f. sucrose
27. On hydrolysis, milk sugar does not yield
    a. fructose   b. galactose   c. glucose
28. On complete hydrolysis, table sugar does not yield
    a. fructose   b. glucose   c. sucrose
29. Which of the following carbohydrates would not yield maltose if it were partially hydrolyzed?
    a. amylose   b. amylopectin   c. cellulose   d. glycogen   e. starch
30. Fructose is also known as
    a. dextrose   b. milk sugar   c. levulose   d. table sugar
31. The mixture of monosaccharides produced by hydrolysis of sucrose is called
    a. blood sugar   b. invert sugar   c. milk sugar   d. table sugar

32. Saccharin is a(n)
    a. hexose   b. aldose   c. reducing sugar   d. glycoside   e. none of these
33. Mannose is a(n)
    a. aldohexose   b. aldopentose   c. ketohexose   d. ketopentose   e. triose
34. 2-Deoxyribose is an isomer of
    a. fructose   b. glyceraldehyde   c. ribose   d. none of these
35. Dihydroxyacetone is an isomer of
    a. glyceraldehyde   b. glucose   c. ribose   d. saccharin   e. none of these
36. Ribose is a(n)
    a. aldopentose   b. ketohexose   c. ketotriose   d. aldotetrose   e. none of these
37. Cellulose is a
    a. monosaccharide   b. disaccharide   c. polysaccharide   d. none of these
38. In which of the following are monosaccharide units not joined by alpha linkages?
    a. maltose   b. amylose   c. glycogen   d. cellulose
39. Which shorthand formula correctly shows the orientation of the secondary hydroxyl groups in D-galactose?

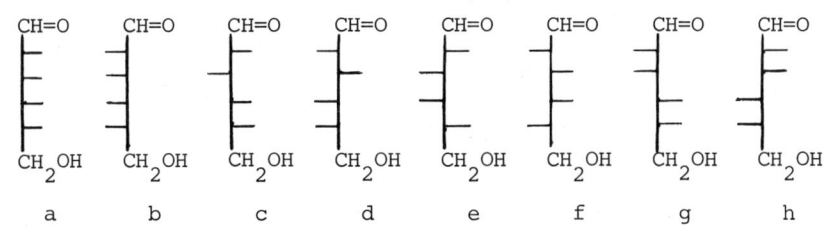

    a   b   c   d   e   f   g   h

40. Which compound contains a beta-acetal linkage?
    a. amylose   b. glycogen   c. lactose   d. maltose   e. starch
41. Which compound does not show 1,6-branching?
    a. amylopectin   b. amylose   c. glycogen
42. Which compound contains alpha acetal linkages and 1,6-branching?
    a. amylose   b. cellulose   c. glycogen   d. sucrose
43. Fructose is not a(n)
    a. alcohol   b. aldehyde   c. carbohydrate   d. sugar   e. saccharide
44. Which is not a reducing sugar?
    a. fructose   b. galactose   c. glucose   d. lactose   e. maltose   f. sucrose
45. If a carbohydrate gives a positive Fehling's test, the carbohydrate
    a. is a reducing sugar
    b. is reduced
    c. will give a negative Tollen's test
46. Which reagent will accomplish the reaction: CH=O → COO$^-$?
    a. Tollen's reagent
    b. Fehling's reagent
    c. Benedict's reagent
    d. All of these reagents will yield the product shown.
    e. None of these reagents will yield the product shown.
47. Which compound will not be oxidized by Fehling's reagent?

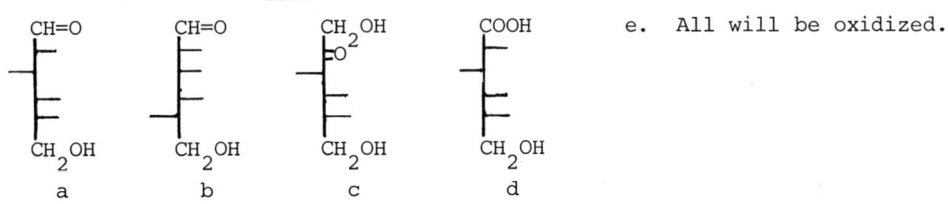

    a   b   c   d   e. All will be oxidized.

48. Is the following compound a reducing sugar?
    a. yes   b. no

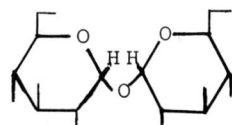

49. Which compound cannot be metabolized by human beings?
    a. amylopectin   b. amylose   c. cellulose   d. maltose   e. starch

50. Which structural feature prevents the digestion of the compound described in question 49.
    a. the branches formed through 1,6-linkages
    b. the presence of galactose units in the polysaccharide
    c. the form of the acetal linkages
    d. the length of the polysaccharide

ANSWERS:

1.	d	26.	c
2.	d	27.	a
3.	a	28.	c
4.	c	29.	c
5.	b	30.	c
6.	e	31.	b
7.	k,l	32.	e
8.	g	33.	a
9.	k	34.	d
10.	g,h,i	35.	a
11.	h	36.	a
12.	j	37.	c
13.	k	38.	d
14.	j,k,l	39.	e
15.	f	40.	c
16.	c	41.	b
17.	h	42.	c
18.	b	43.	b
19.	a	44.	f
20.	a	45.	a
21.	a	46.	d
22.	a	47.	d
23.	c	48.	b
24.	d	49.	c
25.	e	50.	c

Chapter 23 Lipids

The chart below collects all the saponifiable lipids discussed in chapter 23 in one place. The compounds have been arranged to emphasize similarities in the structures and to make it easier to compare the structures.

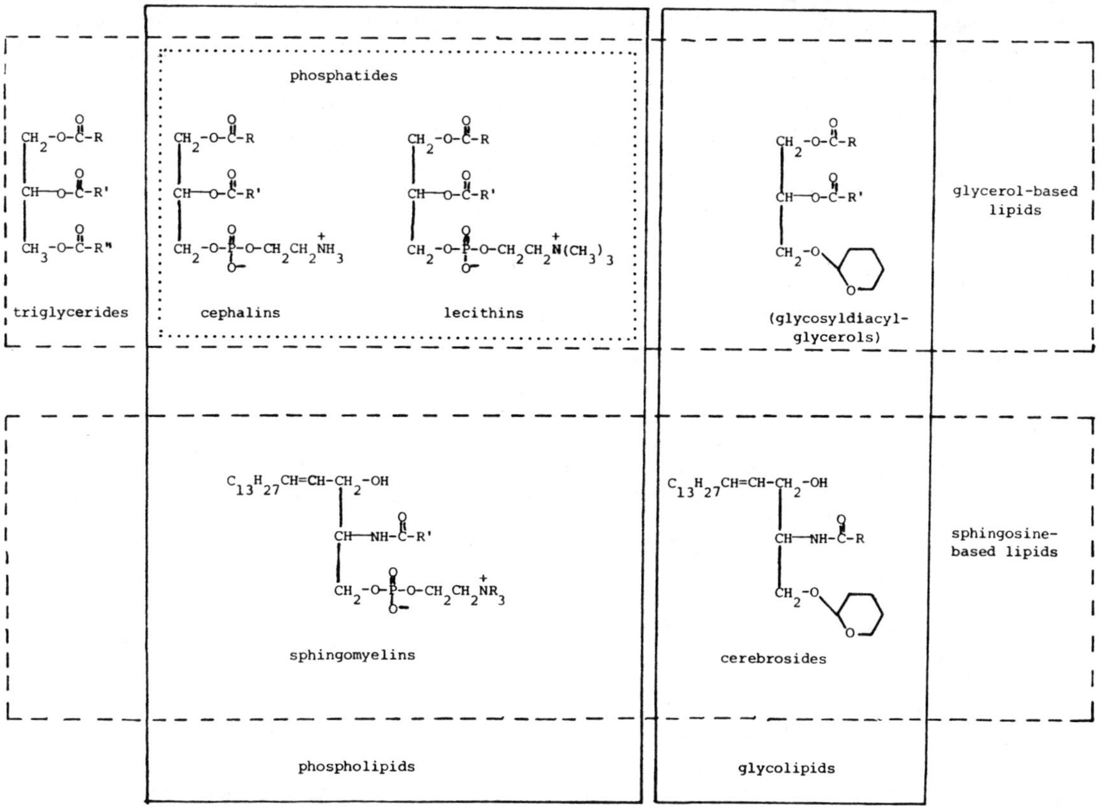

[Chart showing saponifiable lipids organized in a 2×2 grid:

Top-left quadrant (glycerol-based lipids, non-phospholipid): triglycerides structure
Top-middle (phosphatides, within phospholipids region): cephalins and lecithins structures
Top-right quadrant (glycerol-based lipids, glycolipids): glycosyldiacylglycerols structure

Bottom-left (sphingosine-based lipids, phospholipids): sphingomyelins structure
Bottom-right (sphingosine-based lipids, glycolipids): cerebrosides structure

Columns labeled: phospholipids | glycolipids
Rows labeled: glycerol-based lipids | sphingosine-based lipids]

You should be able to draw from memory the triglyceride structure. You should be able to recognize the others. That's not as hard as it sounds, particularly if you have a good grasp of the triglyceride structure. After all--if you know glyco- comes from glycose with means sugar, then identifying a glycolipid should be a snap because we've just spent some time looking at sugars in chapter 22. Identifying a phospholipid should be even easier. Look for phosphorus. The same can be said of sphingosine-based lipids. If you know what the glycerol unit in a triglyceride looks like, then the sphingosine unit in sphingosine-based lipids is obviously different and identifiable.

Note that phospholipid is a more general term than phosphatide. All phosphatides are phospholipids, but not all phospholipids are phosphatides. The phosphatides are those compounds which contain both phosphorus and the glycerol unit.

Names like lecithin and cerebroside give no hint of the corresponding structure (in contrast to designations like phospholipid or glycolipid). Therefore, associating these names with structural features involves some straight memorization. For example, if you want to be able to recognize a cerebroside, you'll have to mentally translate that to "a sphingosine-based glycolipid" and then look for a sphingosine unit and a sugar unit.

The nonsaponifiable steroids are very easy to recognize because of their characteristic fused-ring structure. Actually writing out the structure frequently impresses it much more firmly on one's mind, and we advise you to do this.

Why all this talk about structure? Because as we begin discussing bigger and more complicated molecules, the very appearance of those molecules gets in our way. When such a structure appears on a page, a message seems to travel from eye to brain and says "WARNING...COMPLICATED STRUCTURE...YOU'LL NEVER UNDERSTAND WHAT THEY'RE TALKING ABOUT!" By spending the time to familiarize ourselves with the structures we are shutting off (or at least turning down) that alarm system.

It isn't important that you be able to draw out perfectly the structure of a phospholipid. But you should be able to recognize a significant difference between the phospholipid and a triglyceride. One incorporates an ionized group, the other doesn't. Is that important? Well, one (the triglyceride) is actually stored in the body as droplets of fat within special cells (that's what adipose tissue is--in other words, if you are fat, that's what you've got a lot of). The other, because it has a polar end (the ionized group) and long, nonpolar hydrocarbon chains, can organize its molecules into bilayers and is the structural material of membranes. You can see what very different roles these molecules play in the body because of their structure.

Be sure to go through the problems at the end of the chapter in the text before attempting the self-test.

Self-Test

1. Which is a lipid?
    a. dextrose   b. glycerol   c. tristearin
2. Animal fats and vegetables oils are natural
    a. polymers.   b. soaps   c. esters   d. hydrocarbons   e. carboxylic acids
3. The body's principal energy reserves are in the form of
    a. alcohol   b. carbohydrates   c. enzymes   d. fats   e. proteins
4. The double bond in an unsaturated fatty acid causes a bend in the hydrocarbon chain. Because of this, these compounds, compared to saturated fatty acids,
    a. experience weaker intermolecular attractive forces
    b. have higher melting points
    c. do not react with glycerol
5. Mixed triglycerides contain
    a. at least two different fatty acid units
    b. a phosphate unit and fatty acid units
    c. choline and ethanolamine
6. Oils are
    a. phospholipids   b. liquid fats   c. steroids
7. In general, a vegetable triglyceride would not be hydrogenated to produce
    a. margarine   b. shortening   c. cooking oil
8. The iodine number of a sample of peanut oil is 92; the iodine number of a sample of linseed oil is 198. Which is the more highly saturated oil?
    a. linseed oil   b. peanut oil
9. Hydrogenation of oleic acid yields
    a. lauric acid   b. linoleic acid   c. linolenic acid   d. stearic acid
10. When butter turns rancid, which chemical reaction is not involved?
    a. hydrogenation   b. hydrolysis   c. oxidation
11. Basic hydrolysis of a triglyceride is called
    a. esterification   b. hydrogenation   c. polymerization   d. saponification
12. Sodium salts of long-chain fatty acids are called
    a. lecithins   b. soaps   c. micelles   d. synthetic detergents
13. Which compound would not be expected to exhibit detergent action?
    a. $CH_3CH_2CH_2CH_2CH_2CH_2CH_2CH_2CH_2CH_2CH_2CH_2CH_2CH_2CH_2COO^- Na^+$
    b. $CH_3CH_2CH_2CH_2CH_2CH_2CH_2CH_2CH_2CH_2CH_2CH_2CH_2CH_2COOH$
    c. $CH_3CH_2CH_2CH_2CH_2CH_2CH_2CH_2CH_2CH_2CH_2CH_2CH_2CH_2OSO_3^- Na^+$
14. Which fatty acid salt is soluble in water solutions?
    a. $RCOO^- K^+$ b. $(RCOO^-)_2 Mg^{2+}$ c. $(RCOO^-)_2 Ca^{2+}$ d. none are soluble
15. Complete saponification of a fat yields
    a. fatty acids and glycerol   b. fatty acids and a glyceride
    c. salts of fatty acids and glycerol   d. salts of fatty acids and a glyceride

16. Modern synthetic detergents are not
    a. biodegradable
    b. soluble in hard water
    c. soaps
17. Saponification refers to the reaction of lipids with
    a. an enzyme   b. hydrogen   c. phosphate   d. sodium hydroxide
18. A wax is
    a. a solid fat
    b. any solid lipid
    c. the ester of a long-chain fatty acid and a long-chain alcohol
19. Which is a wax according to the chemical definition of that term?
    a. carbowax
    b. carnauba wax
    c. paraffin wax
20. Spermaceti is a wax obtained from
    a. honeycombs
    b. palm leaves
    c. whales
21. Phosphatides are not
    a. amine-containing lipids
    b. glycerol-based lipids
    c. sphingosine-based lipids
22. Which of the lipids does not incorporate glycerol as part of its structure?
    a. cephalins
    b. lecithins
    c. sphingomyelins
23. Which are glycolipids?
    a. cephalins
    b. cerebrosides
    c. lecithins
24. One distinguishes between a cephalin and a lecithin on the basis of
    a. the amino alcohol unit incorporated in the molecule
    b. the presence or absence of a phosphate group
    c. the presence or absence of a sugar unit
25. Sphingomyelins are lipids based on sphingosine rather than glycerol. In all other respects sphingomelins resemble
    a. glycosides   b. phosphatides   c. triglycerides
26. Which compound is a nonsaponifiable lipid?
    a. cholesterol   b. glyceride   c. phosphatide
27. The bile salts are not
    a. fatty acids   b. emulsifying agents   c. steroids
28. This compound   is:
    a. cholesterol   b. prostaglandin   c. a steroid
29. Cholesterol is not
    a. an alcohol   b. a lipid   c. saponifiable   d. a steroid
30. The prostaglandins are derivatives of
    a. cholesterol   b. a fatty acid   c. glycerol
31. In their physiological action, the prostaglandins resemble
    a. carbohydrates   b. cholesterol   c. enzymes   c. hormones
32. Which special arrangment of polar lipids comes closest to the structure of cell membranes?
    a. bilayers
    b. micelles
    c. monolayers

Answer questions 33 through 40 by referring to the following structure.

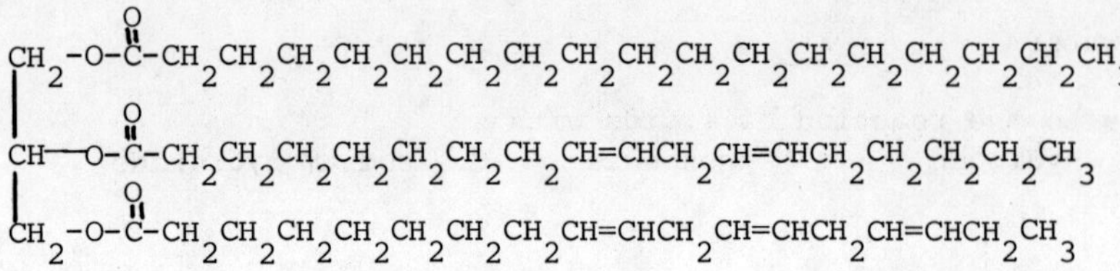

$$CH_2-O-\overset{O}{\underset{\|}{C}}-CH_2CH_2CH_2CH_2CH_2CH_2CH_2CH_2CH_2CH_2CH_2CH_2CH_2CH_2CH_2CH_3$$
$$CH-O-\overset{O}{\underset{\|}{C}}-CH_2CH_2CH_2CH_2CH_2CH_2CH_2CH=CHCH_2CH=CHCH_2CH_2CH_2CH_3$$
$$CH_2-O-\overset{O}{\underset{\|}{C}}-CH_2CH_2CH_2CH_2CH_2CH_2CH_2CH=CHCH_2CH=CHCH_2CH=CHCH_2CH_3$$

33. The compound is a
    a. glyceride   b. phosphatide   c. sphingolipid   d. steroid
34. Hydrolysis of the compound would not yield
    a. choline   b. glycerol   c. stearic acid
35. A high proportion of this type of lipid is found in
    a. animal fats   b. cell membranes   c. vegetable oils
36. The iodine number of a sample of lipid containing a large proportion of this compound would be
    a. high   b. low
37. Saponification of this compound will not yield
    a. salts of glycerol   b. salts of fatty acids   c. soaps
38. The solubility of this compound is similar to that of
    a. glycerol   b. soap   c. cholesterol
39. Compared to a similar but saturated fat, this compound will turn rancid
    a. more rapidly   b. less rapidly
40. If a galactose unit were substituted for the third fatty acid unit, the compound would be a
    a. cephalin   b. cerebroside   c. glycolipid

ANSWERS

1. c	11. d	21. c	31. d
2. c	12. b	22. c	32. a
3. d	13. b	23. b	33. a
4. a	14. a	24. a	34. a
5. a	15. c	25. b	35. c
6. b	16. c	26. a	36. a
7. c	17. d	27. a	37. a
8. b	18. c	28. c	38. c
9. d	19. b	29. c	39. a
10. a	20. c	30. b	40. c

Chapter 24 Amino Acids and Proteins

As has been true of the preceding two chapters, chapter 24 concentrates on structure. The proteins are more comparable in structure to the carbohydrates than to the lipids. Like the polysaccharides, they are polymers. Unlike the polysaccharides we studied, the proteins are copolymers--incorporating more than twenty different monomers. And the monomers are amino acids, not sugars. It is this increased structural complexity which permits proteins to play so many different roles in the body.

You should learn the structures of several representative amino acids. We recommend the following.

Nonpolar Side Chains	Polar Side Chains	Ionizable Side Chains	
		Acidic	Basic
$H_3\overset{+}{N}-CH_2-COO^-$ glycine	$H_3\overset{+}{N}-\underset{\underset{CH_2OH}{\mid}}{CH}-COO^-$ serine	$H_3\overset{+}{N}-\underset{\underset{CH_2COOH}{\mid}}{CH}-COO^-$ aspartic acid	$H_2N-\underset{\underset{(CH_2)_4\overset{+}{N}H_3}{\mid}}{CH}-COO^-$ lysine
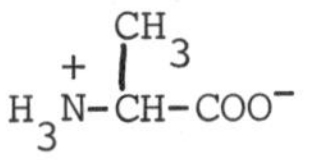 $H_3\overset{+}{N}-\underset{\underset{CH_3}{\mid}}{CH}-COO^-$ alanine	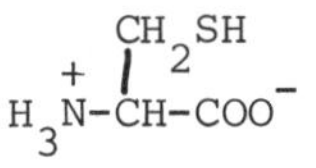 $H_3\overset{+}{N}-\underset{\underset{CH_2SH}{\mid}}{CH}-COO^-$ cysteine (and you should know that cystine is the disulfide formed from cysteine)		
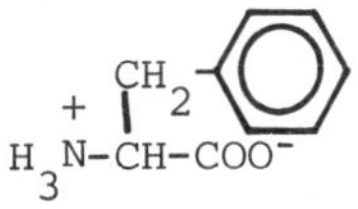 $H_3\overset{+}{N}-\underset{\underset{CH_2-\bigcirc}{\mid}}{CH}-COO^-$ phenylalanine			

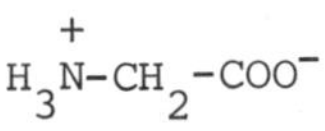

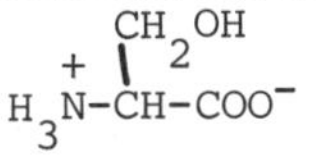

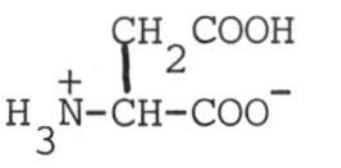

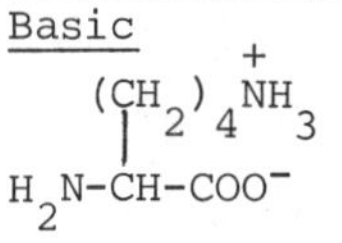

We'll point out what will become obvious as you familiarize yourself with these compounds. Except for glycine and lysine, they can all be considered variations on the alanine structure, that is, each has a functional group attached at the side-chain methyl group of alanine. That fact should make them a bit easier to memorize.

The problems which follow are meant to supplement those in the text. They focus attention on the details of the molecular structure of simple peptides (and their constituent amino acids).

PROBLEMS: 1. For the following structure pick out the peptide bond, a disulfide bond, and an ionizable side chain.

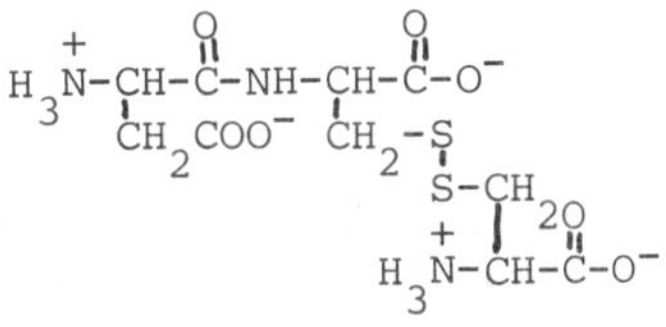

$$H_3\overset{+}{N}-\underset{\underset{CH_2COO^-}{\mid}}{CH}-\overset{O}{\overset{\|}{C}}-NH-\underset{\underset{CH_2-S}{\mid}}{CH}-\overset{O}{\overset{\|}{C}}-O^-$$
$$\underset{\underset{H_3\overset{+}{N}-\underset{\underset{CH_2O}{\mid}}{CH}-\overset{O}{\overset{\|}{C}}-O^-}{}}{S-CH_2}$$

2. Draw the complete structural formulas of two different dipeptides incorporating serine and cysteine.
3. Draw the products of complete hydrolysis of:

$$H_3\overset{+}{N}-\underset{\underset{CH_3}{\mid}}{CH}-\overset{O}{\overset{\|}{C}}-NH-CH_2-\overset{O}{\overset{\|}{C}}-NH-\underset{\underset{CH_2OH}{\mid}}{CH}-\overset{O}{\overset{\|}{C}}-O^-$$

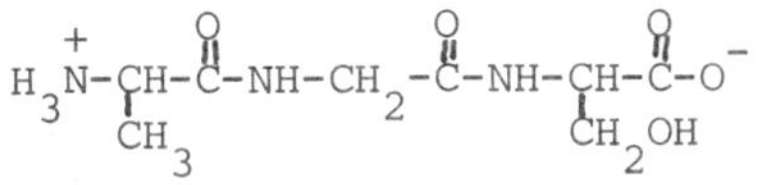

4. Draw the abbreviated formulas (e.g., Gly-Ala-Ser) for all tripeptides which incorporate one unit each of glycine, alanine, and serine.
5. How would the sets of products isolated from complete hydrolysis of each of the tripeptides of problem 4 differ?

One aspect of the chemistry of amino acids and proteins which sometimes causes confusion is the form of the compounds in acid or base. In section 24.4 the structures of a simple amino acid in acidic and basic solutions were given. Let's look at a slightly more complex situation in which the amino acid contains an acidic or basic side chain.

Consider lysine and aspartic acid in a solution of low pH. Remember--low pH means acidic and acidic means there are lots of protons around. If there are lots of protons available, then every group in the amino acid which can carry a proton does. Thus, at low pH aspartic acid and lysine look like this:

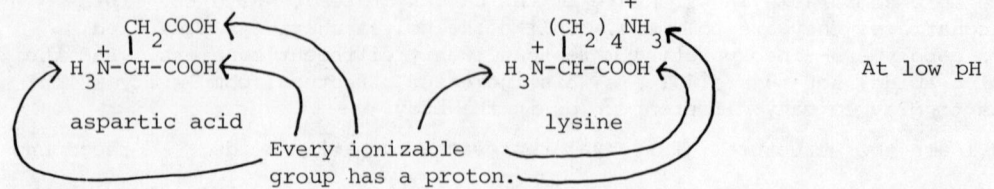

$$\underset{\text{aspartic acid}}{\overset{CH_2COOH}{H_3\overset{+}{N}-CH-COOH}} \qquad \underset{\text{lysine}}{\overset{(CH_2)_4\overset{+}{N}H_3}{H_3\overset{+}{N}-CH-COOH}} \qquad \text{At low pH}$$

Every ionizable group has a proton.

At high pH the solution is basic. Under these conditions, all available protons are plucked from the amino acids.

$$\underset{\text{aspartic acid}}{\overset{CH_2COO^-}{H_2N-CH-COO^-}} \qquad \underset{\text{lysine}}{\overset{(CH_2)_4NH_2}{H_2N-CH-COO^-}} \qquad \text{At high pH}$$

If the pH of a solution is changed from high to low, one after another of the ionizable groups picks up a proton. The strongest bases (the amino groups) react first, then the carboxylate groups (-COO⁻) react. In cases where there are two similar groups present, each picks up a proton at a characteristic pH. From our discussion you have no way of predicting which of the two similar groups reacts first. But we can show you the progressive change for our two model compounds.

Aspartic Acid

$$\overset{CH_2COOH}{H_3\overset{+}{N}-CH-COOH} \underset{H^+}{\overset{OH^-}{\rightleftharpoons}} \overset{CH_2COOH}{H_3\overset{+}{N}-CH-COO^-} \underset{H^+}{\overset{OH^-}{\rightleftharpoons}} \overset{CH_2COO^-}{H_3\overset{+}{N}-CH-COO^-} \underset{H^+}{\overset{OH^-}{\rightleftharpoons}} \overset{CH_2COO^-}{H_2N-CH-COO^-}$$

the zwitterion at the isoelectric point (pH 2.77)

Low pH   High pH

Lysine

$$\overset{(CH_2)_4\overset{+}{N}H_3}{H_3\overset{+}{N}-CH-COOH} \underset{H^+}{\overset{OH^-}{\rightleftharpoons}} \overset{(CH_2)_4\overset{+}{N}H_3}{H_3\overset{+}{N}-CH-COO^-} \underset{H^+}{\overset{OH^-}{\rightleftharpoons}} \overset{(CH_2)_4\overset{+}{N}H_3}{H_2N-CH-COO^-} \underset{H^+}{\overset{OH^-}{\rightleftharpoons}} \overset{(CH_2)_4NH_2}{H_2N-CH-COO^-}$$

the zwitterion at the isoelectric point (pH 9.74)

The same principle applies to the chemistry of polypeptides and proteins.

We emphasize again that you should complete the questions at the end of the chapter in the text before attempting the self-test.

Self-Test

1. Which is not a characteristic element of proteins
    a. Cl   b. C   c. H   d. O   e. N
2. Chemically, proteins are
    a. nucleic acids   b. polyamides   c. polyesters   d. polysaccharides
3. The mirror image isomers of amino acids incorporated in peptides and proteins are members of the
    a. D-family   b. L-family

4. Which is a reasonable representation of a portion of a protein chain?

   a. $-\text{CHCNHCCHNHCNH}-$ b. $-\text{CH}_2\text{CHNHCCH}_2\text{CHNHC}-$ c. $-\text{CHCNHCHCNHCHCNH}-$
       R R   R   R   R   R   R
   (with O above each C=O)

5. Which compound is not an alpha amino acid?

   a. $\text{HO-CH}_2\text{CHCOOH}$ (NH$_2$)
   b. $\text{H}_2\text{N-CH-CH}_3$ (COOH)
   c. $\text{H}_2\text{N-CH-CH}_2\text{COOH}$ (CH$_3$)
   d. H-C-NH_2 (H, COOH)

6. $\overset{+}{\text{H}_3}\text{N-CH-CONH-CH}_2\text{CONH-CH-COO}^-$ (with CH$_3$ and CH$_2$OH side chains) is a(n):

   a. amino acid   b. tripeptide   c. dipeptide   d. polypeptide   e. protein

7. Which is not considered evidence of the zwitterionic structure of amino acids?
   a. They show greater solubility in water than in nonpolar solvents.
   b. They have high decomposition points.
   c. They can polymerize to form proteins.

8. Amino acid side chains do not include
   a. hydrocarbon groups   b. ionized groups   c. polar groups
   d. a disulfide bond   e. phosphate esters

9. Glycine is

   a. $^+\text{NH}_3\text{-CH}_2\text{COO}^-$
   b. $^+\text{NH}_3\text{-CH}_3\text{CHCOO}^-$
   c. $^+\text{NH}_3\text{-CH}_2\text{CH}_2\text{COO}^-$
   d. $\text{SH-CH}_2\text{-CHCOO}^-$ ($^+\text{NH}_3$)

10. Which amino acid contains a sulfhydryl group?
    a. cysteine   b. leucine   c. lysine   d. serine

11. γ-Aminobutyric acid is
    a. an essential amino acid
    b. the principal amino acid incorporated in the protein of silk
    c. a chemical neurotransmitter found in the brain

12. To indicate that the order of a segment of peptide is Lys-Gly-Ala-Cys is to describe its __?__ structure.
    a. primary   b. secondary   c. tertiary   d. quaternary

13. The primary structure of a protein is dependent on the formation of
    a. disulfide bonds   b. hydrogen bonds   c. peptide bonds   d. salt bridges

14. If we say a protein exists as an alpha helix, we are describing its __?__ structure.
    a. primary   b. secondary   c. tertiary   d. quaternary

15. Which amino acid side chain will participate in hydrophobic interactions to maintain the tertiary structure of a protein?
    a. aspartic acid   b. lysine   c. phenylalanine

16. By describing the relative position of the four polypeptide chains of the hemoglobin molecule, we specify its __?__ structure.
    a. primary   b. secondary   c. tertiary   d. quaternary

17. The pleated sheet conformation refers to the __?__ structure of a protein.
    a. primary   b. secondary   c. tertiary   d. quaternary

18. The structure of collagen can be described as a(n)
    a. alpha helix   b. double helix   c. triple helix   d. pleated sheet

19. Which term is not used in describing protein structure?
    a. alpha helix   b. double helix   c. pleated sheet   d. triple helix

20. Which food source fails to provide all of the essential amino acids?
    a. corn   b. eggs   c. fish   d. meat   e. milk

21. Which of the following terms cannot be used to describe the structure of alanine at its isoelectric point?
    a. dipolar ion   b. electrically neutral   c. inner salt   d. zwitterion
    e. peptide

22. In a strongly basic solution, which form will predominate?

    a. $\text{CH}_3\text{-CH-COOH}$ (NH$_2$)
    b. $\text{CH}_3\text{-CH-COO}^-$ ($^+\text{NH}_3$)
    c. $\text{CH}_3\text{-CH-COO}^-$ (NH$_2$)
    d. $\text{CH}_3\text{-CH-COOH}$ ($^+\text{NH}_3$)

23. In a solution of very low pH, aspartic acid molecules would exist as

    a. $\text{H}_2\text{N-CH-COOH}$ (CH$_2$COOH)
    b. $\text{H}_2\text{N-CH-COO}^-$ (CH$_2$COO$^-$)
    c. $\text{H}_3\overset{+}{\text{N}}\text{-CH-COO}^-$ (CH$_2$COOH)
    d. $\text{H}_3\overset{+}{\text{N}}\text{-CH-COO}^-$ (CH$_2$COO$^-$)
    e. $\text{H}_3\overset{+}{\text{N}}\text{-CH-COOH}$ (CH$_2$COOH)

24. At the isoelectric point of proteins
    a. the proteins are least soluble
    b. the proteins contain no charged groups
    c. the proteins have a large excess of positive charge
    d. the pH of the solution is always 7
25. Oxytocin and vasopressin are polypeptide
    a. enzymes   b. hormones   c. structural material
26. Which is not a globular protein?
    a. albumin   b. collagen   c. myoglobin
27. Which compound does not contain heme as a prosthetic group?
    a. albumin   b. hemoglobin   c. myoglobin
28. Which of the following processes is least likely to have occurred during the denaturation of a protein?
    a. disruption of hydrogen bonds   b. hydrolysis of peptide bonds
    c. cleavage of disulfide bonds   d. disruption of salt bridges
29. Which is not a denaturing agent?
    a. heat   b. CH_3CH_2OH c. Hg^{2+} d. alkaloidal reagents   e. H_2O
30. Which type of protein is more easily denatured?
    a. fibrous   b. globular

True or False

T F 31. An essential amino acid is one which must be incorporated in every protein.
T F 32. The most energy-efficient diet is one rich in meat products.
T F 33. The most abundant amino acid in silk protein is glycine.
T F 34. Proteins and peptides are arbitrarily distinguished by molecular weight.
T F 35. Wool fibers are considerably more elastic than silk fibers because the secondary structure of wool protein is alpha helical.
T F 36. The amino acid sequence in hemoglobin shows no variation with species.
T F 37. The peptide bonds in proteins are identical to the bonds which link monomer units in the nylon polymer.
T F 38. In the Van Slyke analysis, proteins are treated with ninhydrin and produce a purple color.
T F 39. Fibrous proteins are more soluble in aqueous solutions than are globular proteins.
T F 40. The hydrogen bonds formed between different peptide linkages play a major role in establishing both the pleated sheet and the alpha helix conformations in proteins.

ANSWERS

Problems:
1. [structure showing peptide bond, ionizable side chain, disulfide bond]
2. [two tripeptide structures with CH_2OH, CH_2SH groups]
3. [three amino acid structures: alanine, glycine, serine]
4. Gly-Ala-Ser; Gly-Ser-Ala; Ala-Gly-Ser; Ala-Ser-Gly; Ser-Gly-Ala; Ser-Ala-Gly
5. There would be no difference. Hydrolysis of each tripeptide would yield a mixture of glycine, alanine and serine.

Self-Test:
1. a   9. a   17. b   25. b   33. T
2. b   10. a   18. c   26. b   34. T
3. b   11. c   19. b   27. a   35. T
4. c   12. a   20. a   28. b   36. F
5. c   13. c   21. e   29. e   37. T
6. b   14. b   22. c   30. b   38. F
7. c   15. c   23. e   31. F   39. F
8. e   16. d   24. a   32. F   40. T

Chapter 25 Nucleic Acids

The best review of chapter 25 involves working the problems at the end of the chapter. We shall add little more to that review before presenting the self-test.

In the self-test, we will expect you to be able to recognize the distinguishing features of nucleic acids, nucleotides and nucleosides. We'll also expect you to recognize which type of compound is being discussed from its name (study table 25.2). Although you will not be asked to draw the complete structure of the various bases, you should recognize a purine when you see one (two fused heterocyclic rings) and a pyrimidine (one heterocyclic ring). You should also know which bases are purines (adenine and guanine) and which are pyrimidines (cytosine, thymine and uracil).

To give you a warmup before the self-test, try the following problems which review other points covered in the chapter. In each labelled drawing, there is an error. You are being asked to spot the error.

1. A typical nucleoside

2. A nucleotide obtained from the hydrolysis of DNA

3. A typical base pair in a nucleic acid

4. Beginning of DNA replication

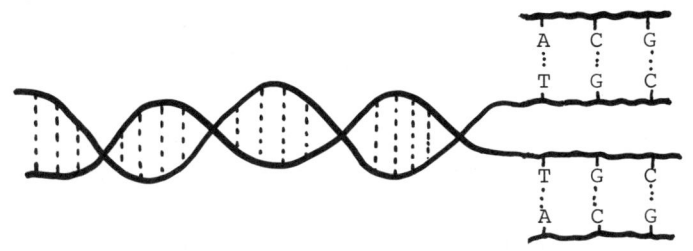

5. DNA serving as a template for the formation of a m-RNA molecule

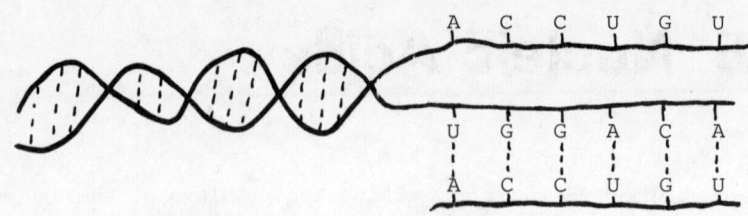

6. Formation of a protein molecule at the ribosomes

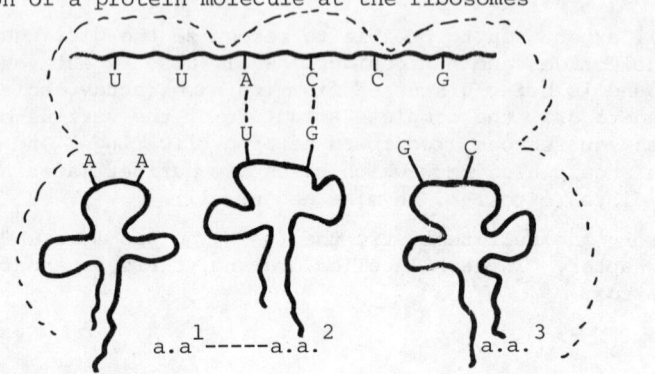

Self-Test
1. A gene is a segment of a molecule of
   a. DNA   b. mRNA   c. tRNA   d. protein
2. If a nucleic acid is completely hydrolyzed, which type of compound is not one of the products?
   a. a purine   b. a pyrimidine   c. phosphoric acid   d. an amino acid
   e. a sugar
3. Which set of bases does not make up a base pair usually found in nucleic acids?
   a. adenine-thymine   b. cytosine-guanine   c. uracil-thymine   d. adenine-uracil
4. The organic molecules which constitute most genes incorporate
   a. ribose   b. deoxyribose
5. Which group is not part of the backbone of a strand of nucleic acid?
   a. sugar unit   b. base unit   c. phosphoric acid unit
6. Base pairing is accomplished through the formation of
   a. hydrogen bonds   b. phosphate linkages   c. hemiacetal linkages
7. Which base is not found in RNA?
   a. adenine   b. cytosine   c. guanine   d. thymine   e. uracil
8. Which molecule is a nucleoside?
   a. cytosine   b. cytidine   c. cytidine monophosphate   d. deoxycytidine monophosphate
9. Which nucleotide would be incorporated in a DNA molecule?
   a. guanine   b. adenosine   c. guanosine monophosphate   d. deoxyadenosine monophosphate
10. The base sequence CGA would not pair with
    a. GCU   b. GCT   c. AGC
11. Which contains the codon?
    a. DNA   b. mRNA   c. tRNA   d. the protein molecule
12. Which molecule carries the anticodon?
    a. mRNA   b. tRNA   c. the ribosome   d. the protein molecule
13. The genetic message is transcribed from DNA to
    a. mRNA   b. tRNA   c. DNA
14. Where does gene replication take place?
    a. in a cell nucleus
    b. at a ribosome complex
    c. at the cell membrane
15. Where does the transcription of information from DNA to mRNA take place?
    a. in a cell nucleus
    b. at a ribosome complex
    c. at the cell membrane

16. After replication, each daughter DNA molecule
    a. contains only purines or pyrimidines, but not both.
    b. contains one strand of the parent molecule.
    c. is the mirror image isomer of the other daughter molecule.
17. Which process occurs at the ribosome complex?
    a. replication of DNA   b. transcription to mRNA   c. translation to protein
18. When active protein synthesis is taking place in the cell, which material is not required at the ribosomes?
    a. DNA   b. mRNA   c. tRNA   d. growing protein chain
19. If CCG/ser represents a tRNA molecule with amino acid attached, which of the following is not a possible tRNA molecule.
    a. GGC/ser   b. GCC/gly   c. CCG/gly   d. AGC/ser
20.

    a. AMP   b. cyclic AMP   c. ADP   d. ATP

Answer questions 21 through 26 by referring to the following structure.

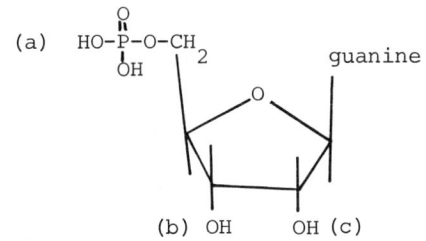

21. The compound is a
    a. nucleoside   b. nucleotide   c. nucleic acid
22. The compound is
    a. guanine   b. guanine monophosphate   c. deoxyguanine   d. guanosine monophosphate
    e. deoxyguanosine monophosphate
23. The compound incorporates a
    a. purine   b. pyrimidine
24. The compound could be incorporated in
    a. DNA   b. RNA
25. Three of the -OH groups have been labelled a, b, and c. Which one of these would not be used in the formation of the nucleic acid polymer?
    a   b   c
26. If the compound were incorporated in a nucleic acid, which base would not appear in the same polymer?
    a. adenine   b. cytosine   c. guanine   d. thymine   e. uracil

True or False

T F 27. It is impossible for base pairing to occur in single-stranded RNA.
T F 28. Nucleic acids, polysaccharides and proteins are all polymers.
T F 29. The molecular weights of nucleic acids are generally greater than those of proteins.
T F 30. DNA always incorporates equal amounts of purines and pyrimidines.
T F 31. In nucleoproteins the basic side chains of the protein form salt bridges with the base pairs of the nucleic acids.
T F 32. Nucleotides are formed in the hydrolysis of nucleosides.
T F 33. Adenylic acid is identical to adenosine monophosphate.
T F 34. It is the presence of the ribose unit in the nucleic acid RNA which makes the compound an acid.

T F 35. The pairing of a purine with a pyrimidine permits the strands of a double helix to maintain a constant spacing.
T F 36. The pairing of cytosine with guanine and adenine with thymine permits the best hydrogen-bonded interactions between base pairs.
T F 37. Transfer RNA contains both the codon triplet and the amino acid called for by the codon.
T F 38. The codons which do not call for a specific amino acid signal the termination of protein synthesis.
T F 39. Thymine and uracil are both pyrimidines.
T F 40. Some codons call for more than one kind of amino acid.

ANSWERS

Problems:
1. The base uracil is shown attached to a deoxyribose ring. Uracil is only found with ribose.
2. The sugar unit is ribose. DNA would yield only deoxyribose.
3. Both bases are purines. A typical base pair would include a purine and a pyrimidine.
4. The bases on the complementary strands of the DNA molecule are not complementary. T was paired with T, G with G, etc. In DNA a strand carrying T and G and C would be matched with one carrying A and C and G.
5. Among the bases attached to the double stranded DNA molecule is uracil. This base is only found in RNA.
6. The tRNAs are pairing with doublets rather than the correct triplets.

Self-Test:

1.	a	11.	b	21.	b	31.	F
2.	d	12.	b	22.	d	32.	F
3.	c	13.	a	23.	a	33.	T
4.	b	14.	a	24.	b	34.	F
5.	b	15.	a	25.	c	35.	T
6.	a	16.	b	26.	d	36.	T
7.	d	17.	c	27.	F	37.	F
8.	b	18.	a	28.	T	38.	T
9.	d	19.	c	29.	T	39.	T
10.	c	20.	c	30.	T	40.	F

Chapter 26 Enzymes and Coenzymes

The specific enzymes mentioned in chapter 26 are presented below, arranged in alphabetical order for easy reference. The section numbers listed with each enzyme refer to sections in other chapters of this text in which reactions involving the enzymes are discussed.

Enzymes Considered in Chapter 26

acetylcholinesterase--a hydrolase which catalyzes the formation and cleavage of acetylcholine, the compound involved in transmission of nerve impulses across synapses (section 20.11)

aminopeptidase--a hydrolase found in intestinal juices; it catalyzes the cleavage of the terminal peptide bond at the free amino end of a protein chain (section 29.4)

arginase--a hydrolase found in mammalian liver; it catalyzes the conversion of L-arginine to ornithine and urea; this reaction is one of a series that is responsible for removing nitrogen wastes from the body (section 32.5)

carboxypeptidase--a hydrolase produced in the form of its proenzyme by the pancreas and secreted into the small intestine where it is activated; this protease catalyzes the cleavage of the terminal peptide bond at the free carboxyl end of a protein chain (section 29.5)

chymotrypsin--a hydrolase which is found as its proenzyme in pancreatic juice secreted into the small intestine; when activated, this protease usually catalyzes the cleavage of peptide bonds involving amino acids with aromatic side chains (section 29.5)

creatine phosphokinase (creatine kinase)--an enzyme which catalyzes the transfer of a phosphate group between creatine and ADP; the reaction is involved in the transfer of energy during muscle contraction (sections 30.8 and 32.4)

enterokinase--another of the enzymes found in the intestinal juice; it is a protease responsible for the cleavage of a hexapeptide group from the free amino end of the trypsinogen chain; this reaction converts the proenzyme to the active enzyme trypsin (sections 29.4 and 29.5)

factor X_a--the proteolytic enzyme responsible for the conversion of prothrombin to thrombin; this reaction is one of the series associated with the mechanism of blood clotting (discussed in section 26.10 only)

glutamic-oxalacetic transaminase--a transferase found in a variety of tissues; it catalyzes the transfer of an amino group between a keto acid and an amino acid (section 32.4)

lipase--a general term for those hydrolytic enzymes which catalyze the cleavage of ester bonds in lipids; the digestive process in humans involves both gastric lipase (in the stomach) and pancreatic lipase (in the small intestine) (sections 29.3 and 29.5)

lysozyme--a hydrolytic enzyme that catalyzes the cleavage of acetal linkages in polysaccharides incorporated in bacterial cell walls (section 28.10)

maltase--another hydrolase found in intestinal juice; maltase is the carbohydrase responsible for the cleavage of the disaccharide maltose to the monosaccharide glucose (section 29.4)

papain--isolated from papaya, this hydrolase does not have a very high specificity and catalyzes the cleavage of peptide bonds, ester linkages, etc. (discussed in section 26.10 only)

pepsin--the major protease found in the stomach (section 29.3)

ptyalin (salivary amylase)--a carbohydrase found in the mouth; ptyalin catalyzes the initial hydrolysis of starch in the digestive process (section 29.2)

thrombin--a protease which cleaves the peptide bond between arginine and glycine in fibrinogen and thus converts soluble fibrinogen to the insoluble, clot-forming substance, fibrin (discussed in section 26.10 only) trypsin--this hydrolase, secreted by the pancreas as the proenzyme trypsinogen, is activated in the small intestine where it catalyzes the cleavage of peptide bonds involving the amino acids arginine and lysine (sections 29.4 and 29.5)

urease--a highly specific hydrolase which catalyzes the conversion of urea to carbon dioxide and ammonia (discussed in section 26.3 only)

zymase--the name given to a collection of enzymes isolated from yeast; these enzymes catalyze the fermentation of glucose to ethanol and carbon dioxide (section 16.4)

In section 26.2 we presented the classification of enzymes by types of reactions catalyzed. The best way to illustrate what we mean by type of reaction is to write an equation. Therefore, we present below an example for each of the six general types of enzymes. Remember that these examples are representative. We could have written many others, some seemingly quite different. We chose to use examples of reactions you have already or will soon encounter in this text.

Types of Enzymes with Examples

I. Hydrolases - example was discussed in section 20.11

$(CH_3)_3\overset{+}{N}-CH_2CH_2-O-\overset{O}{\underset{\|}{C}}-CH_3 + H_2O \xrightleftharpoons{\text{acetylcholinesterase}} (CH_3)_3\overset{+}{N}-CH_2CH_2-OH + HO-\overset{O}{\underset{\|}{C}}-CH_3$

acetylcholine   choline   acetic acid

II. Oxidases (oxidoreductases) - example discussed in section 30.4

$CH_3-\underset{\underset{\text{lactic acid}}{}}{\overset{OH}{\underset{|}{CH}}}-COOH + NAD^+ \xrightleftharpoons{\text{lactate dehydrogenase}} CH_3-\overset{O}{\underset{\|}{C}}-COOH + NADH + H^+$

lactic acid   pyruvic acid

III. Transferases - example discussed in section 30.4

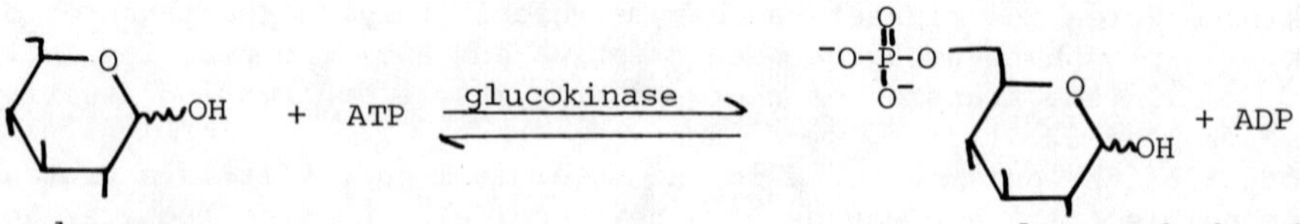

glucose + ATP $\xrightleftharpoons{\text{glucokinase}}$ glucose-6-phosphate + ADP

IV. Lyases - example discussed in section 30.4

$CH_3-\overset{O}{\underset{\|}{C}}-COOH \xrightleftharpoons{\text{pyruvate decarboxylase}} CH_3-\overset{O}{\underset{\|}{C}}-H + CO_2$

pyruvic acid   acetaldehyde

V. Isomerases - example discussed in section 30.1

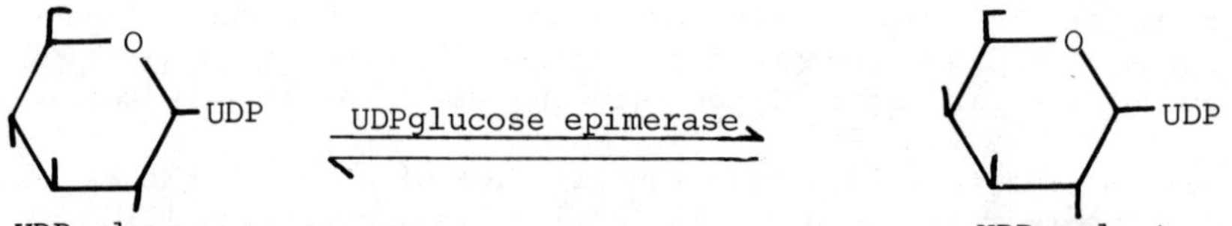

UDP-glucose $\xrightleftharpoons{\text{UDPglucose epimerase}}$ UDP-galactose

VI. Ligases - example discussed in section 31.3

$CoA-SH + HO-\overset{O}{\underset{\|}{C}}-R + ATP \xrightleftharpoons{\text{acyl-CoA synthetase}} CoA-S-\overset{O}{\underset{\|}{C}}-R + AMP + P_i$

fatty acid   thio ester of fatty acid

Answer the problems at the end of the chapter in the text, then complete the self-test.

Self-Test
1. A compound which catalyzes a chemical reaction in a living organism is called a(n)
   a. carbohydrate   b. enzyme   c. lipid   d. vitamin
2. Enzymes belong to which class of organic compounds?
   a. carbohydrates   b. esters   c. hydrocarbons   d. lipids   e. proteins
3. The compound whose reaction is catalyzed by an enzyme is called a(n)
   a. activator   b. coenzyme   c. cofactor   d. substrate
4. Which type of enzyme is not classified as a hydrolase?
   a. carbohydrase   b. dehydrogenase   c. lipase   d. peptidase   e. protease
5. Kinases belong to which class of enzymes?
   a. hydrolases   b. isomerases   c. ligases   d. lyases   e. oxidases
   f. transferases
6. To which of the classes listed in question 5 does a synthetase belong?
   a   b   c   d   e   f
7. Which type of enzyme listed in question 5 would catalyze the conversion of L-alanine to D-alanine?
   a   b   c   d   e   f
8. Since thrombin catalyzes only the cleavage of the peptide bond ···Arg-Gly···, it is said to be
   a. group specific   b. linkage specific   c. reaction specific   d. stereospecific
9. An apoenzyme is always a(n)
   a. protein   b. non-protein organic molecule   c. inorganic ion
10. Which is not found as a cofactor of an enzyme?
    a. a protein   b. a vitamin   c. an inorganic ion
11. A vitamin which is required for the functioning of an enzyme system is called a(n)
    a. activator   b. apoenzyme   c. coenzyme
12. When all the active sites on enzyme molecules are saturated, an increase in __?__ concentration will not increase the rate of the reaction.
    a. enzyme   b. substrate
13. Enzymes operating in the stomach should have an optimum pH in the __?__ range.
    a. acidic   b. neutral   c. basic
14. The optimum temperature for most enzymes operating in the human body is
    a. 273 K   b. 37 °C   c. 98.6 °C   d. 100 °C
15. Which is not a term used to describe an enzyme which controls the rate of reactions governed by a set of enzymes?
    a. allosteric   b. autocatalytic   c. regulatory   d. all terms are correct
16. The allosteric site on a regulatory enzyme binds the
    a. inhibitor   b. proenzyme   c. substrate
17. Which is not a proenzyme?
    a. pepsinogen   b. prothrombin   c. trypsinogen   d. all are proenzymes
18. Pepsin is a protease which acts in the
    a. mouth   b. stomach   c. intestines
19. Which material is involved in blood clotting?
    a. fibrinogen   b. pepsinogen   c. trypsinogen
20. Which is not a poison which affects enzymes?
    a. CN^- b. Fe^{3+} c. Pb^{2+} d. AsO_4^{3-}
21. Which poison does not act by tying up sulfhydryl groups of an enzyme?
    a. CN^- b. AsO_4^{3-} c. Pb^{2+}
22. EDTA is an effective antidote for acute poisoning by
    a. CN^- b. AsO_4^{3-} c. Pb^{2+}
23. In one form of hemophilia, the following enzyme is missing.
    a. CPK   b. Factor X_a c. Fibrin
24. Cyanide acts as a poison by
    a. interrupting cell respiration
    b. blocking oxidation of glucose in the cell
    c. tying up electrons required for the reduction of oxygen in the cell
    d. All of the above are true.
25. Activation of an enzyme can occur through
    a. conversion of a proenzyme to the enzyme
    b. combination of an apoenzyme with a cofactor
    c. release of an inhibitor from the allosteric site of a regulatory enzyme
    d. All of the above are true.

True or False

T F 26. Enzymes are heat stable catalysts.
T F 27. Factors that affect the activity of an enzyme include temperature, pH, enzyme concentration and substrate concentration.
T F 28. Saliva contains the enzyme ptyalin.
T F 29. The active site of an enzyme is the point at which an inhibitor attaches.
T F 30. The term "regulatory enzyme" is used for any enzyme which catalyzes a chemical reaction.
T F 31. Conversion of a proenzyme to the active enzyme frequently involves cleavage of a portion of the protein chain.
T F 32. The carbohydrase lysozyme was the first enzyme whose three-dimensional structure was determined by scientists.
T F 33. A stereospecific enzyme would catalyze only reactions involving both members of a mirror image pair of isomers.
T F 34. If an enzyme exhibits absolute specificity, it catalyzes a single reaction.
T F 35. The formation of an activated enzyme-substrate complex is postulated to explain the catalytic effect of an enzyme.
T F 36. Addition of an enzyme changes the speed and the position of the equilibrium of a reaction.
T F 37. The lock and key theory states that the substrate must fit precisely the active site of the enzyme for the catalyst to be most effective.
T F 38. The amino acid side chains associated with the active site of an enzyme must be adjacent in the primary sequence of the protein.
T F 39. The inhibitor of an allosteric enzyme is frequently the end product of the sequence of reactions controlled by the enzymes.
T F 40. CPK and GOT are abbreviations for the enzymes used in detergent formulations and meat tenderizers.

The enzymes listed in column A can act as catalysts for the reactions illustrated in column B. For each reaction, indicate which enzyme serves as catalyst.

Column A

a. carbohydrase
b. dehydrogenase
c. isomerase
d. peptidase
e. transferase

Column B

_____ 41. ~C(=O)-NH-CH$_2$~ + H$_2$O ⇌ ~C(=O)-OH + H$_2$N-CH$_2$~

_____ 42. H$_3$N$^+$-C(CH$_3$)(H)-COO$^-$ ⇌ $^-$OOC-C(CH$_3$)(H)-NH$_3^+$

_____ 43. (disaccharide) + H$_2$O ⇌ (monosaccharide) + (monosaccharide)

_____ 44. CH$_3$-CH$_2$-OH ⇌ CH$_3$-CH(=O) + 2H

_____ 45. CH$_3$-C(=O)-OH + ATP ⇌ CH$_3$-C(=O)-O-P(=O)(O$^-$)-O$^-$ + ADP

ANSWERS

1.	b	11.	c	21.	a	31.	T	41.	d
2.	e	12.	b	22.	c	32.	T	42.	c
3.	d	13.	a	23.	b	33.	F	43.	a
4.	b	14.	b	24.	d	34.	T	44.	b
5.	f	15.	b	25.	d	35.	T	45.	e
6.	c	16.	a	26.	F	36.	F		
7.	b	17.	d	27.	T	37.	T		
8.	b	18.	b	28.	T	38.	F		
9.	a	19.	a	29.	F	39.	T		
10.	a	20.	b	30.	F	40.	F		

Chapter 27 Vitamins and Hormones

Chapter 27, like the preceding ones, deals with complex molecules. However, there is no structural feature common to all vitamins and hormones. Proteins are polyamides and it is possible to describe in a general way structural features common to thousands of different protein molecules. The same can be said about carbohydrates and even about lipids. But neither vitamins nor hormones can be so easily categorized by structure. Because many of these compounds have such complex structures, they frequently intimidate students (who, because of exams, worry a great deal about being able to draw things like vitamin B_{12}). Therefore, let us first comment that there is not one chemist in a hundred (or more) who can draw the structure of vitamin B_{12} from memory.

So why did we bother showing the structures for all of these molecules? Because we want you to see that molecular architecture is more than a chemist's playground; it frequently determines the state of one's health. Consider the discussion of vitamin A in section 27.2. This is a large, relatively nonpolar molecule. We've discussed polarity, hydrogen bonding, solubility, etc., many times. Here you have a concrete example of the significance of this "chemistry." Vitamin A is nonpolar and soluble in nonpolar media like the fatty tissue of the body. It does not form hydrogen bonds with water in sufficient numbers to make it soluble in aqueous body fluids. That means that it is not rapidly excreted from the body with such fluids. Thus, individuals can build up a reserve and protect themselves against the effects of deprivation. It is also true that this same property of vitamin A permits one to overdose on the vitamin. That just means you can store too much of it--the body doesn't automatically dump the excess over one's immediate needs.

For contrast, look at vitamin C. It's a relative of the carbohydrates--lots of hydroxyl groups, very polar and very water-soluble. Now, perhaps, you can understand some of the controversy surrounding Linus Pauling's recommendation about massive doses of vitamin C. What difference does it make, say some scientists, whether you ingest 250 or 15 000 mg of vitamin C when evidence suggests that the aqueous body fluids wash out all but 200 mg?

With these facts in mind, spend some time just looking at (not memorizing) some of the vitamin structures. Notice that vitamins A, D, E and K, the fat-soluble vitamins, are all lipid-like. They have lots of carbon and hydrogen and very little else. There's an occasional oxygen, but mostly there are long, nonpolar chains of carbon.

Now look at the B complex vitamins (table 27.1). All of them contain nitrogen. Even more important, however, is the fact that they all contain a relatively high proportion of groups which can interact through hydrogen bonding. The variation in structure among the B vitamins is great, but B_2 (riboflavin) is typical. Like most of the fat-soluble vitamins, it has a side chain. But look at the riboflavin side chain. It carries 4 hydroxyl groups--and there are four nitrogens and two other oxygens in the molecule. Look at vitamin B_{12} (cyanocobalamine). The laboratory synthesis of this complex molecule was regarded as one of the outstanding achievements of organic chemistry in this century. What should you know about the structure? Certainly that it contains cobalt, which is somewhat unusual, but primarily that it is loaded with groups that confer water-solubility --amides and other nitrogen-containing functions, hydroxyl groups and a phosphate group.

The hormones also vary considerably in structure, but those listed in table 27.3 are either proteins, amino acid derivatives or steroids. Of the sex hormones discussed in this chapter, perhaps the most notable feature is the extraordinary similarity of compounds which elicit such distinctive responses in our bodies.

You should use problem 3 at the end of the chapter in the text to organize for yourself some pertinent data for each of the vitamins. Go through all of the problems in the text before working the self-test.

Self-Test - Select the best answer.

1. Vitamins are
   a. amines required by an organism for good health
   b. organic compounds produced in trace amounts by the endocrine glands
   c. organic molecules which an organism requires in trace amounts but cannot synthesize for itself
   d. steroids which act as sex hormones
2. Which is not a fat-soluble vitamin?
   a. vitamin A b. vitamin C c. vitamin D d. vitamin E e. vitamin K
3. Which is not a water-soluble vitamin?
   a. ascorbic acid b. cyanocobalamine c. retinol d. vitamin B_6
4. Which is not a member of the B complex?
   a. ascorbic acid b. biotin c. folic acid d. niacin e. pantothenic acid
   f. riboflavin g. thiamin
5. The plant pigment named β-carotene is a(n)
   a. coenzyme b. contraceptive c. hormone d. provitamin e. vitamin
6. A critical event in the chemistry of vision involves the conversion of
   a. a cis isomer to a trans isomer
   b. a D isomer to an L isomer
   c. an ortho isomer to a para isomer
   d. ergosterol to calciferol
7. Which is not true of rhodopsin?
   a. It is a complex of a protein and a derivative of vitamin A.
   b. It is the visual pigment found in some receptor cells of the retina.
   c. It is converted to vitamin A by the absorption of light.
8. Which is not true of vitamin D?
   a. It is formed from steroidal precursors by the absorption of ultraviolet light.
   b. It is called the "sunshine vitamin".
   c. A deficiency of this vitamin results in abnormal bone formation.
   d. No harmful effects have been documented for overdoses of this vitamin.
9. Vitamin E is
   a. an antioxidant
   b. frequently missing from the diet of vegetarians
   c. approved by medical authorities for the prevention of aging
10. The vitamin associated with blood clotting is vitamin
    a. A b. B complex c. C d. D e. E f. K
11. The B complex vitamins are frequently incorporated in
    a. coenzymes b. provitamins c. contraceptives d. rhodopsin
12. The B complex vitamins are characterized by
    a. long hydrocarbon side chains
    b. a relatively high proportion of polar functional groups
    c. a tendency to hydrogen bond with fats in the body
13. The B complex vitamins can be stored in almost unlimited quantities in the
    a. adipose tissue b. bone marrow c. liver d. retina
    e. They are not stored in significant amounts in the body.
14. Vitamin C activity is exhibited by
    a. several pigments isolated from various colored plants
    b. steroid-like compounds found in the skin of various animals
    c. a carbohydrate-like compound found in citrus fruit
15. In megavitamin therapy
    a. water-soluble vitamins are prescribed in sufficient amounts to permit storage in the liver
    b. placebos are substituted for vitamin C tablets
    c. vitamin dosages are tailored to individual requirements
16. Which gland is responsible for the production of releasing factors which trigger the pituitary gland?
    a. adrenal cortex b. hypophysis c. hypothalamus
17. Hormones function as
    a. chemical messengers b. coenzymes c. provitamins
18. Which are male sex hormones?
    a. androgens b. estrogens c. progestins

19. An effective oral contraceptive was produced when two structural features were combined in one compound. Which two?
    a. the combination of a male sex hormone with a female sex hormone
    b. the removal of a methyl group from and addition of an ethynyl group to an appropriate steroid
    c. the combination of a trans double bond and an aldehyde group
20. Which component of the typical birth-control pill is responsible for regulating the menstrual cycle?
    a. androgen   b. estrogen   c. progestin
21. Which is not true of DES?
    a. It is a synthetic female sex hormone.
    b. It is banned as an additive in cattle feed.
    c. It is banned as an abortifacient.

True or False
T F 22. Cortisone is a vitamin which exhibits antiinflammatory properties.
T F 23. All hormones are steroids synthesized in trace amounts by the endocrine glands.
T F 24. Hormones are secreted by endocrine glands located in the target organs.
T F 25. Male sex hormones have been used to treat breast cancer in women.
T F 26. The presence of estrogen in contraceptive pills is thought to be responsible for most of the undersirable side effects of these pills.

Matching
Match the names in column B with the vitamins in column A.

Column A
____ 27. Vitamin A
____ 28. Vitamin B$_1$
____ 29. Vitamin B$_2$
____ 30. Vitamin B$_{12}$
____ 31. Vitamin C
____ 32. Vitamin D
____ 33. Vitamin E

Column B
a. ascorbic acid
b. calciferol
c. cyanocobalamine
d. retinol
e. riboflavin
f. thiamine
g. alpha-tocopherol

Match the deficiency diseases or symptoms in column D with the vitamins in column C.

Column C
____ 34. ascorbic acid
____ 35. cyanocobalamine
____ 36. niacin
____ 37. thiamine
____ 38. vitamin D
____ 39. vitamin E
____ 40. vitamin K

Column D
a. beriberi
b. hemorrhage
c. pellagra
d. pernicious anemia
e. rickets
f. scurvy
g. sterility

ANSWERS
1. c   11. a   21. c   31. a
2. b   12. b   22. F   32. b
3. c   13. e   23. F   33. g
4. a   14. c   24. F   34. f
5. d   15. c   25. T   35. d
6. a   16. c   26. T   36. c
7. c   17. a   27. d   37. a
8. d   18. a   28. f   38. e
9. a   19. b   29. e   39. g
10. f   20. b   30. c   40. b

Chapter 28 Body Fluids

Chapter 28 represents a change of pace. For some time now we have been examining various classes of biochemically important molecules. Our survey of these important compounds is now complete, and we are embarking on an examination of how these materials function in the body. In this chapter we've begun that examination by considering the properties of many of the fluids in which biochemical reactions take place. So in this chapter there's very little that's new as far as molecules are concerned. Hemoglobin is the only compound whose structure is considered in any detail, and even hemoglobin has been encountered previously (section 24.12).

Much of the material in the chapter, therefore, was descriptive. Many applications of chemical principles in living systems were provided. For the most part, this chapter does not require that you learn new scientific principles, but rather that you relate familiar chemistry to biological systems.

The only numerical problems associated with this chapter deal with concentrations of electrolytes. Problem 8 at the end of the chapter tests your understanding of this material. Since it has been some time since we've encountered numerical problems, we'll add a few more practice exercises here.

I. Normal values for various electrolytes in urine collected over a 24-hour period are:
- calcium (Ca^{2+}): 2.5 - 20 meq
- chloride (Cl^+): 110 - 250 meq
- magnesium (Mg^{2+}): 6.0 - 8.5 meq
- potassium (K^+): 40 - 80 meq
- sodium (Na^+): 80 - 180 meq

Analysis of a 24-hour urine sample detected the following amounts of electrolytes. Indicate for each ion whether the amount is normal or abnormal.
- calcium: 100 mg
- chloride: 3.55 g
- magnesium: 4.86 mg
- potassium: 1.955 g
- sodium: 6.9 g

II. Normal concentration ranges for various electrolytes in blood serum are:
- calcium: 4.6 - 5.5 meq/ℓ
- chloride: 98 - 110 meq/ℓ
- magnesium: 1.3 - 2.1 meq/ℓ
- potassium: 3.6 - 5.5 meq/ℓ
- sodium: 135 - 155 meq/ℓ

The following concentrations were reported for a sample of serum. In each case indicate whether the value falls within the normal range.
- calcium: 4 mg%
- chloride: 355 mg%
- magnesium: 2.43 mg%
- potassium: 39.1 mg%
- sodium: 345 mg%

There is only one other thing we wish to clarify before presenting the self-test--the transportation of carbon dioxide by the blood. If you flip through the pages of the chapter, you'll notice this is the only point at which we use chemical equations to any extent. In that sense, it is more chemical and less biological than other parts of the chapter. And since our job is to teach the chemistry, let us summarize briefly the sequence of steps involved in carbon dioxide transport.

In Capillaries Serving Metabolically-Active Tissue

Carbon dioxide produced as a metabolic product in tissue cells migrates into the erythrocytes where it combines with water.

$$CO_2 + H_2O \longrightarrow H_2CO_3$$

The resulting carbonic acid protonates the conjugate base of the hemoglobin buffer in the erythrocyte.

$$H_2CO_3 + Hb^- \longrightarrow HCO_3^- + HHb$$

The resulting bicarbonate ion dissolves in the fluid within the erythrocyte and in the surrounding blood plasma. Any bicarbonate ion which migrates from the erythrocyte to the surrounding plasma is matched by the migration of a chloride ion from the surrounding plasma into the erythrocyte (chloride shift).

In Capillaries Serving the Lungs

The conjugate acid of the hemoglobin buffer picks up oxygen to become the conjugate acid of the oxyhemoglobin buffer.

$$HHb + O_2 \longrightarrow HHbO_2$$

The conjugate acid of the oxyhemoglobin buffer protonates the bicarbonate ion.

$$HHbO_2 + HCO_3^- \longrightarrow HbO_2^- + H_2CO_3$$

The resulting carbonic acid dissociates into water and carbon dioxide.

$$H_2CO_3 \longrightarrow H_2O + CO_2$$

The carbon dioxide gas is exhausted to the atmosphere.

Self-Test
1. Which material does blood serum not include?
   a. electrolytes   b. fibrinogen   c. proteins
2. Which of the plasma proteins are associated with the immune response of the body?
   a. albumin   b. fibrinogen   c. globulins   d. prothrombin
3. Which of the plasma proteins contributes most to the oncotic pressure of blood?
   a. albumin   b. fibrinogen   c. globulin   d. prothrombin
4. Which of the cations is not one of the principal electrolytes in blood plasma?
   a. Na^+ b. K^+ c. Ca^{2+} d. Mg^{2+} e. Fe^{2+}
5. Which of the anions is not one of the principal electrolytes in blood plasma?
   a. Cl^- b. HCO_3^- c. CO_3^{2-} d. HPO_4^{2-} e. SO_4^{2-}
6. Material moves in and out of the capillaries primarily through the process of
   a. diffusion   b. evaporation   c. filtration
7. Which ion is necessary for blood clotting?
   a. Cl^- b. HCO_3^- c. Ca^{2+} d. Fe^{2+}
8. Which ion is not part of the blood buffers?
   a. HCO_3^- b. HPO_4^{2-} c. HSO_4^-
9. Oncotic pressure refers to
   a. the pressure imparted to the blood by the pumping action of the heart
   b. the osmotic pressure of the blood due to dissolved electrolytes
   c. the osmotic pressure of the blood due to proteins present in colloidal dispersion
10. The hydrostatic pressure of the blood is
    a. higher at the venous end of a capillary
    b. higher at the arterial end of a capillary
    c. approximately the same at both ends of a capillary
11. Which describes the conditions at the arterial end of the capillaries?
    a. oncotic pressure exceeds hydrostatic pressure and the net movement of fluid is into the interstitial compartment
    b. oncotic pressure exceeds hydrostatic pressure and the net movement of fluid is into the capillary
    c. hydrostatic pressure exceeds oncotic pressure and the net movement of fluid is into the interstitial compartment
    d. hydrostatic pressure exceeds oncotic pressure and the net movement of fluid is into the capillary
12. In edema, there is a net flow of fluid to the interstitial compartment because
    a. the oncotic pressure of the blood increases
    b. the oncotic pressure of the blood decreases
    c. the hydrostatic pressure of the blood decreases
13. Which is not characteristic of shock?
    a. fluid accumulates in the vascular system

b. capillary permeability increases and the protein concentration in interstitial fluid increases
   c. the blood pressure decreases
   d. all of the above are characteristic of shock
14. An antibody is
   a. a foreign macromolecule which triggers the body's immune response
   b. a protein formed by the body to attack specific foreign particles
   c. a pathogenic microorganism
15. A vaccine contains
   a. a weakened antigen
   b. a weakened antibody
   c. gamma globulin
16. Carbon dioxide produced in metabolic reactions is carried to the lungs chiefly as
   a. free CO_2 gas   b. dissolved CO_2 gas   c. HCO_3^- d. a prosthetic group of hemoglobin
17. Which reaction occurs in the capillaries of the lungs?
   a. $HHbO_2 \longrightarrow HHb + O_2$
   b. $HHbO_2 + HCO_3^- \longrightarrow HbO_2^- + H_2CO_3$
   c. $H_2CO_3 + Hb^- \longrightarrow HHb + HCO_3^-$
18. Which condition will cause the indicated shift in equilibrium for the bicarbonate/carbonic acid buffer: $H^+ + HCO_3^- \longleftarrow H_2CO_3 \longleftarrow H_2O + CO_2$?
   a. buildup of lactic acid during vigorous exercise
   b. inadequate ventilation due to emphysema
   c. hyperventilation
19. To maintain electrical neutrality as bicarbonate ion diffuses out of erythrocytes
   a. potassium ions accompany the bicarbonate ions
   b. sodium ions accompany the bicarbonate ions
   c. chloride ions diffuse into the erythrocyte
20. Which is not true of a hemoglobin molecule?
   a. It is a conjugated protein.
   b. It incorporates iron in a +2 oxidation state.
   c. It contains four identical protein chains in a roughly tetrahedral arrangement.
   d. It incorporates four heme units, each of which can bind with an oxygen molecule.
21. Bilirubin is a product of
   a. the breakdown of the prosthetic group of hemoglobin
   b. the breakdown of the alpha chain of hemoglobin
   c. the breakdown of the beta chain of hemoglobin
22. The abnormal hemoglobin which is characteristic of sickle cell anemia contains
   a. iron in the +3 oxidation state
   b. a heme unit containing no iron ion
   c. peptide chains incorporating an incorrect amino acid
23. The constitution of the fluid within the lymphatic system is identical to
   a. blood plasma
   b. interstitial fluid
   c. intracellular fluid
24. Lymph nodes are not involved in the manufacture of
   a. antibodies   b. erythrocytes   c. leukocytes
25. Which type of nutrient is absorbed into the lymphatic system from the intestine?
   a. carbohydrate   b. fat   c. protein
26. Salts of which cation are not ordinarily found in kidney stones?
   a. Na^+ b. Ca^{2+} c. Mg^{2+}
27. Which substance is not filtered out of the blood at the glomerulus?
   a. erythrocytes   b. glucose   c. urea   d. water
28. Which substance is least likely to be reabsorbed in the kidney tubules?
   a. amino acids   b. glucose   c. urea   d. water
29. Insensible perspiration is the water lost
   a. through the respiratory tract
   b. from the sweat glands
   c. from the lacrimal glands

30. Galactosemia is a condition which results from
   a. an inability to metabolize the breakdown products from milk sugar
   b. a faulty genetic code which produces hemoglobin S
   c. the rejection of transplanted tissue by the immune response

True or False

T	F	31.	Heat is one of the major metabolic products carried off by perspiration.
T	F	32.	Lysozyme is a type of bacteria occasionally found in lacrimal fluid.
T	F	33.	Casein is the outer oily layer of tears.
T	F	34.	Mammals living in cold climates produce milk with high fat content.
T	F	35.	Polycythemia and anemia are both conditions associated with abnormal concentrations of red blood cells.
T	F	36.	Plasma can be isolated if an anticoagulant is first added to freshly drawn blood.
T	F	37.	Sodium ions are found mainly in the plasma and potassium ions are found mainly in the erythrocytes.
T	F	38.	Calcium and phosphorus levels in the blood usually vary reciprocally.
T	F	39.	Metabolic disorders more commonly produce the condition known as alkalosis rather than acidosis.
T	F	40.	The chloride shift refers to the loss of chloride ion to the urine when its threshold level is exceeded.

ANSWERS

Problems:
I. calcium = 5 meq, normal
   chloride = 100 meq, slightly low
   magnesium = 0.4 meq, very low
   potassium = 50 meq, normal
   sodium = 300 meq, very high

II. calcium = 2 meq/ℓ, low
    chloride = 100 meq/ℓ, normal
    magnesium = 2 meq/ℓ, normal
    potassium = 10 meq/ℓ, high
    sodium = 150 meq/ℓ, normal

Self-Test:
1. b
2. c
3. a
4. e
5. c
6. a
7. c
8. c
9. c
10. b
11. c
12. b
13. a
14. b
15. a
16. c
17. b
18. b
19. c
20. c
21. a
22. c
23. b
24. b
25. b
26. a
27. a
28. c
29. a
30. a
31. T
32. F
33. F
34. T
35. T
36. T
37. T
38. T
39. F
40. F

Chapter 29 Digestion

Chapter 29 continues the pattern established in the preceding chapter. Having considered the makeup of various fluids in the body in chapter 28, we've now turned our attention to fluids that function in the tunnel running through the body and, therefore, formally classified as outside of the body. As was the case in the preceding chapter, most of the material in chapter 29 is descriptive. Carbohydrates, fats and proteins (including enzymes) were discussed several chapters back. So, too, were their hydrolysis products. All we've done in this chapter is to show you where some digestive enzymes meet their substrates--and what happens to the substrates we call foods as they chug through the digestive tract. We've simply described the chemistry of digestion in its biological setting.

Thus, there is not much explaining to do about material in this chapter. The questions at the end of the chapter will take you through a review of the material. The following table offers a summary of the process of digestion. It presumes you've studied the chapter and simply condenses much of the data into a compact form for easy reference.

Summary of Digestive Process

Digestive Organ	Digestive Fluid	Enzyme Present	Nutrients Acted On	Other Important Compounds Present
Mouth	Saliva	ptyalin	carbohydrates	mucin - lubricates masticated food
Stomach	gastic juice	pepsinogen ↓ HCl, pepsin ↓ pepsin	proteins	histamine and the hormone gastrin - start flow of gastric juice hydrochloric acid - denatures proteins and antivates pepsin intrinsic factor - necessary for absorption of vitamin B_{12}
		gastric lipase	fats (limited)	
Small intestine	intestinal juice	sucrase, maltase, lactase	carbohydrates	the hormone secretin - triggers production of pancreatic juice the hormone cholecystokinin - triggers the emptying of the gall bladder
		aminopeptidase, dipeptidases, tripeptidases	proteins, peptides	
		nucleotidases	nucleic acids	
	pancreatic juice	trypsinogen ↓ enterokinase trypsin chymotrypsinogen ↓ trypsin chymotrypsin procarboxypeptidase ↓ trypsin carboxypeptidase	proteins	bicarbonate ion - to neutralize chyme and establish optimum pH for enzymes
		steapsin	fats	
		pancreatic amylase	carbohydrates	
	bile		fats	bicarbonate ion - to raise pH sodium taurocholate and sodium glycocholate - bile salts which emulsify fats
Large Intestine (primarily functions to remove water from the feces)			carbohydrates amino acids	bacterial colonies

Self-Test

1. The chief chemical reaction in the gastrointestinal tract is
    a. hydration   b. hydrolysis   c. oxidation   d. reduction
2. One of the fluids does not contain enzymes. Which one?
    a. bile   b. gastric juices   c. intestinal juice   d. pancreatic juice   e. saliva

3. The pancreas is responsible for the production of
   a. biotin   b. insulin   c. norepinephrine
4. Mucin accomplishes the following:
   a. hydrolysis of proteins to amino acids
   b. hydrolysis of polysaccharides to monosaccharides
   c. lubrication of food particles for easy swallowing
   d. emulsification of fats for better contact with enzymes
5. α-Amylase appears in the digestive juices which are active in the
   a. mouth and stomach   b. mouth and small intestine   c. stomach and small intestine
6. Which organ does not contribute digestive juices to the small intestine?
   a. kidney   b. liver   c. pancreas   d. small intestine
7. Which constituent of bile precipitates as gallstones?
   a. bile acids   b. bile pigments   c. cholesterol
8. Which ion acts to neutralize the chyme as it enters the small intestine?
   a. Ca^{2+} b. HCO_3^- c. HPO_4^{2-} d. Cl^- e. Na^+
9. The salivary glands produce an enzyme which initiates digestion of
   a. carbohydrates   b. lipids   c. nucleic acids   d. proteins
10. Which is not a bile salt?
    a. sodium bicarbonate   b. sodium glycocholate   c. sodium taurocholate
11. Bile salts do not aid in the absorption of
    a. cholesterol   b. fats   c. nucleic acids   d. vitamins A,D,E and K
12. Which enzyme does not catalyze the hydrolysis of peptides or proteins?
    a. pepsin   b. steapsin   c. trypsin
13. Which enzyme does not catalyze the hydrolysis of carbohydrates?
    a. alpha-amylase   b. enterokinase   c. lactase   d. ptyalin
14. Which enzyme is not secreted in inactive form as a proenzyme?
    a. alpha-amylase   b. carboxypeptidase   c. chymotrypsin   d. pepsin   e. trypsin
15. Which enzyme does not act in the small intestine?
    a. aminopeptidase   b. maltase   c. ptyalin   d. trypsin
16. Which compound does not catalyze the conversion of pepsinogen to pepsin?
    a. hydrochloric acid   b. pepsin   c. pepsinogen
17. Intrinsic factor is secreted in the
    a. mouth   b. stomach   c. small intestine   d. large intestine

For each of the following indicate which of the listed hormones triggers the action described.

18. the flow of gastric juice
    a. cholecystokinin   b. gastrin   c. insulin   d. secretin
19. the secretion of pancreatic juice
    a. cholecystokinin   b. gastrin   c. insulin   d. secretin
20. the emptying of the contents of the gallbladder
    a. cholecystokinin   b. gastrin   c. insulin   d. secretin

Digestion of each of the classes of food proceeds in two of the three listed digestive organs. For each class of food, indicate where digestion does not take place.

21. carbohydrates: a. mouth   b. stomach   c. small intestine
22. fats: a. mouth   b. stomach   c. small intestine
23. proteins: a. mouth   b. stomach   c. small intestine

24. Which is not true of bacteria in the large intestine?
    a. They convert carbohydrates to organic acids and intestinal gases.
    b. They convert amino acids to odorous compounds.
    c. When released into the abdominal cavity, they produce the vitamin biotin.
25. The feces typically contain
    a. bacteria   b. cellulose   c. water   d. all of these

True or False

T F 26. The large intestine is the site of the most intense digestive activity.
T F 27. Histamine is an enzyme found in the mouth.
T F 28. Digestion is the process by which food materials are converted into substances which can be absorbed and used by the body.
T F 29. Lipase catalyzes the hydrolysis of lactose.

T F 30. The end products of protein digestion are peptides.
T F 31. Carbohydrates are absorbed into the blood as monosaccharides.
T F 32. Digestion of carbohydrates in the mouth produces dextrins.
T F 33. Chyme is an enzyme which catalyzes the hydrolysis of carbohydrates.
T F 34. The pH of the stomach is high.
T F 35. Both hydrochloric acid and the bile acids act to improve contact between enzymes and their substrates.
T F 36. Intrinsic factor is required for the absorption of vitamin B_{12}.
T F 37. The villi are projections which line the stomach and produce hydrochloric acid.
T F 38. Fats are absorbed from the intestine into the lymphatic system.
T F 39. Stomach ulcers may be irritated by gastric juices.
T F 40. The organic acids produced in the colon by bacterial action stimulate the muscular action of the colon.
T F 41. Putrescine and cadaverine are produced by decarboxylation of basic amino acids in the colon.
T F 42. The main function of the colon is the reabsorption of water.
T F 43. Normally the stool is colored by reduction products of bilirubin.
T F 44. A black stool may indicate bleeding in the upper digestive tract or the presence of high levels of iron compounds in the diet.
T F 45. Diets rich in fibers, peels and similar nondigestible materials are usually considered unhealthy.

ANSWERS:

1.	b	10.	a	19.	d	28.	T	37.	F
2.	a	11.	c	20.	a	29.	F	38.	T
3.	b	12.	b	21.	b	30.	F	39.	T
4.	c	13.	b	22.	a	31.	T	40.	T
5.	b	14.	a	23.	a	32.	T	41.	T
6.	a	15.	c	24.	c	33.	F	42.	T
7.	c	16.	c	25.	d	34.	F	43.	T
8.	b	17.	b	26.	F	35.	T	44.	T
9.	a	18.	b	27.	F	36.	T	45.	F

Chapter 30 Carbohydrate Metabolism

We've referred to metabolism, metabolites, and metabolic products frequently in past chapters. In this chapter, we are taking our first extended look at metabolic processes in human beings. The single most noticeable feature of metabolic reactions is that transformations which can be summarized in one equation usually proceed by mechanisms which involve many steps. The very complexity of these processes makes life possible, but it also makes studying the processes difficult.

Let's take a second look at the Embden-Meyerhof pathway and the Krebs cycle and note overall patterns which might make each of these metabolic pathways more easy to comprehend. First, remember both of these pathways are designed to produce energy, which means ATP molecules. Second, there is a most obvious difference between the two series of reactions --one is cyclic (the Krebs cycle) and the other is not (the Embden-Meyerhof pathway). The latter starts with glucose and ends with lactic acid. The former starts and ends with oxaloacetic acid.

Now let's be a little more specific and focus on the Embden-Meyerhof pathway for the moment (figure 30.8). In essense, here's what happens: phosphate groups are added to sugar molecules, which split in two and pick up more phosphate until, finally, a high-energy phosphate (1,3-diphosphoglyceric acid) is formed. The beauty of this compound lies in its ability to transfer a phosphate group to ADP. That's what it does. After the transfer, the remaining compound rearranges a bit to become another high-energy phosphate, PEP. PEP is also able to transfer phosphate to ADP. That leaves pyruvic acid, which can be reduced to lactic acid (the end product of the Embden-Meyerhof pathway) or fed into the Krebs cycle. In the Embden-Meyerhof pathway, a sugar derivative is oxidized at step 6 and NAD^+ is reduced to NADH; but in step 11, a sugar derivative is reduced and NADH is oxidized to NAD^+. Thus, there is no net oxidation or reduction in the Embden-Meyerhof pathway.

Here's the Embden-Meyerhof pathway showing only glucose and its products. See if the comments give you a sense of the direction of the reactions, a sense of a grand design in which everything is done for a purpose. Also--you might again notice how the sacrifice of two ATP molecules prepares the way for the synthesis of four ATP molecules. (These systems are not unaware of an old rule of business--you must sometimes spend money to make money.)

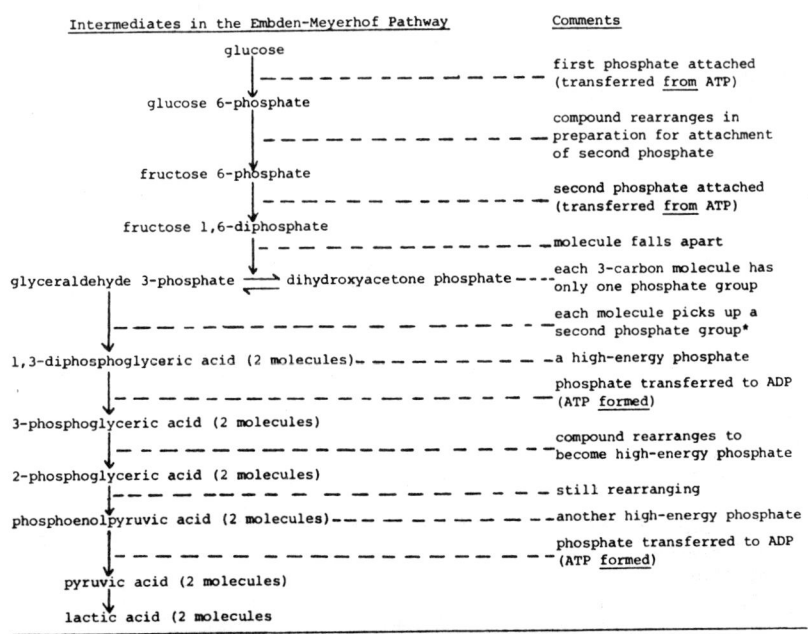

Intermediates in the Embden-Meyerhof Pathway	Comments
glucose	first phosphate attached (transferred from ATP)
glucose 6-phosphate	compound rearranges in preparation for attachment of second phosphate
fructose 6-phosphate	second phosphate attached (transferred from ATP)
fructose 1,6-diphosphate	molecule falls apart
glyceraldehyde 3-phosphate ⇌ dihydroxyacetone phosphate	each 3-carbon molecule has only one phosphate group
	each molecule picks up a second phosphate group*
1,3-diphosphoglyceric acid (2 molecules)	a high-energy phosphate
	phosphate transferred to ADP (ATP formed)
3-phosphoglyceric acid (2 molecules)	compound rearranges to become high-energy phosphate
2-phosphoglyceric acid (2 molecules)	still rearranging
phosphoenolpyruvic acid (2 molecules)	another high-energy phosphate
	phosphate transferred to ADP (ATP formed)
pyruvic acid (2 molecules)	
lactic acid (2 molecules)	

*This is a most important step. These molecules pick up phosphate, but not at the expense of an ATP molecule.

How about the Krebs cycle? Can we make any sense of the pattern of reactions? Well, we can certainly try. What the Krebs cycle does, in essence, is oxidize acetic acid to two carbon dioxide molecules. A chemist can do that in a laboratory by burning acetic acid (that is, by carrying out the combustion of acetic acid). Cells, of course, are far more subtle.

The Krebs cycle (figure 30.9) starts by taking the acetic acid (activated by attachment to Coenzyme A) and bonding it to one of the cycling compounds (oxaloacetic acid). The resulting product is manipulated to produce a nifty little organic molecule which is especially good at decarboxylating. By manipulated we mean that water is removed, then replaced in a different position, then hydrogen is removed. The resulting product does just what is is supposed to do--it decarboxylates. The product from that reaction decarboxylates again--and there go the two carbon dioxide molecules we wanted to produce. Now all we have to do is manipulate the product a bit to get back to where we started. This time the molecule is manipulated as follows: hydrogen is removed, water is added, and some more hydrogen is removed. And--ta da!--there's good old oxaloacetic acid again. What we'd like you to see is the reasonableness of the process. If a cell can not set fire to acetic acid to achieve its ends, then it simply builds molecules which will accomplish the same results.

While there was no net oxidation or reduction in the Embden-Meyerhof pathway, oxidation is the name of the game in the Krebs cycle. At four different points in the cycle (steps 4, 6, 8 and 10) oxidation occurs. The oxidizing agents required for these steps are generated by the electron transport system which operates in oxidative phosphorylation. It is here that ATP is actually synthesized in reactions coupled to the transport of electrons.

In addition to the metabolic fate of glucose, chapter 30 considers the regulation of blood glucose levels and the energetics of muscle contraction. Problems 1 through 5 at the end of the chapter review the first topic and problem 18 reviews the second.

We can't resist making one more comment before we present the self-test. Next time you eat some candy or hear a paramedic told to "start an i.v. with D5W", think about the chain of reactions which is required to utilize that sugar. You might also consider how fortunate we are that our cells are better at chemistry than the best chemist who ever lived.

Self-Test

1. Which is not a path followed by glucose in the body?
    a. conversion to glycogen for storage
    b. conversion to fat for storage
    c. oxidation to produce energy
    d. Glucose follows all of the above paths.
2. An overdose of insulin produces a condition called
    a. galactocemia   b. hyperglycemia   c. hypoglycemia
3. Which hormone triggers a decrease in blood sugar levels?
    a. cortisone   b. glucagon   c. insulin   d. epinephrine
    e. human growth hormone
4. Which is not a pancreatic hormone?
    a. adrenalin   b. glucagon   c. insulin
5. Which compound does glycogen most closely resemble?
    a. amylopectin   b. amylose   c. cellulose   d. glucose
6. Monosaccharides can all be converted to a common intermediate which can be converted to glycogen. What is the intermediate?
    a. acetyl coenzyme A   b. fructose 6-phosphate   c. glucose 1-phosphate   d. pyruvic acid
7. Infants suffering from galactosemia lack the enzyme required for which transformation?
    a. lactose $\rightleftarrows$ glucose and galactose
    b. UDP-galactose $\rightleftarrows$ UDP-glucose
    c. galactose $\rightleftarrows$ galactose 1-phosphate
8. Which enzyme does not catalyze the lengthening of a glycogen chain?
    a. glycogen synthetase
    b. branching enzyme
    c. phosphorylase

9. The Cori cycle describes
    a. the relationship between glycogenesis and glycogenolysis
    b. the conversion of acetic acid to carbon dioxide
    c. the interconversion of monosaccharides
10. The Embden-Meyerhof pathway is also known as
    a. anaerobic glycolysis   b. glyconeogenesis   c. Krebs cycle
    d. oxidative phosphorylation
11. Which compound serves as the intermediate through which all of the monosaccharides enter the Embden-Meyerhof pathway?
    a. acetyl coenzyme A   b. fructose 6-phosphate   c. glucose 1-phosphate   d. pyruvic acid
12. In which sequence are phosphate groups transferred from the high-energy phosphates PEP and 1,3-diphosphoglyceric acid to ADP?
    a. Embden-Meyerhof pathway
    b. Krebs cycle
    c. oxidative phosphorylation
13. What is the end product of anaerobic glycolysis?
    a. acetic acid   b. ethyl alcohol   c. glycogen   d. lactic acid
14. Which sequence of reactions most closely parallels that found in fermentation?
    a. Embden-Meyerhof pathway
    b. Krebs cycle
    c. respiratory chain
15. Which is not another name for the Krebs cycle?
    a. citric acid cycle   b. Cori cycle   c. tricarboxylic acid cycle
16. Through what intermediate do pyruvic acid and lactic acid enter the Krebs cycle?
    a. acetyl coenzyme A   b. ADP   c. fructose 6-phosphate   d. NAD^+
17. Which oxidizing agent is used in both anaerobic glycolysis and in the Krebs cycle?
    a. FAD   b. NAD^+ c. O_2
18. In which sequence of reactions is CO_2 produced?
    a. anaerobic glycolysis   b. Krebs cycle   c. oxidative phosphorylation
19. In which sequence of reactions is ATP not formed?
    a. anaerobic glycolysis   b. Krebs cycle   c. oxidative phosphorylation
20. Which is not one of the intermediates in the Krebs cycle?
    a. citric acid   b. isocitric acid   c. fumaric acid   d. lactic acid   e. succinic acid
21. Two molecules undergo decarboxylation in the Krebs cycle. Which does not?
    a. α-ketoglutaric acid   b. oxaloacetic acid   c. oxalosuccinic acid
22. What process does not occur in oxidative phosphorylation?
    a. conversion of oxygen to water
    b. synthesis of the oxidizing agents NAD^+ and FAD
    c. synthesis of ATP from ADP
    d. oxidation of lactic acid to pyruvic acid
    e. All of the above occur in oxidative phosphorylation.
23. Which is not true of the cytochromes?
    a. They are iron-containing proteins.
    b. They participate in the series of reactions called the respiratory chain.
    c. Their action is strongly inhibited by carbon dioxide.
    d. All of the above are true of the cytochromes.
24. Which high-energy phosphate serves as the immediate source of energy in muscle contraction?
    a. ATP   b. creatine phosphate   c. PEP
25. Actomyosin is
    a. the enzyme which catalyzes the transfer of phosphate from creatine phosphate to ADP
    b. the protein complex which constitutes the contractile tissue of muscle
    c. the drug used to counter the effects of an accumulated oxygen debt

True or False

T F 26. Glucose is stored in the body in the form of glycogen.
T F 27. The glucose tolerance test is used to diagnose galactosemia.
T F 28. Oral drugs used to treat mild diabetes are synthetic forms of insulin.
T F 29. Emotional glycosuria results from epinephrine-stimulated glycogen breakdown.
T F 30. The branching enzyme transfers a polysaccharide chain from a C-4 attachment to a C-6 position.
T F 31. Creatine phosphate is one of the high-energy phosphates produced in the Embden-Meyerhof pathway.

T F 32. Cytochromes are proteins involved in electron transport.
T F 33. There are no redox reactions in the Embden-Meyerhof pathway.
T F 34. Glycogen is known as blood sugar.
T F 35. Fluoroacetic acid acts as a poison by disrupting the Krebs cycle.
T F 36. Cyanide compounds act as poisons by disrupting the Embden-Meyerhof pathway.
T F 37. The iron in cytochromes may be in the +2 or +3 oxidation state.
T F 38. In periods of normal (not strenuous) activity, energy is supplied to muscles through aerobic pathways.
T F 39. Creatine phosphate serves as a readily available energy reserve in muscle.
T F 40. While the Embden-Meyerhof pathway supplies the ATP required for muscular activity, the body incurs an oxygen debt.

ANSWERS:

1.	d	11.	b	21.	b	31.	F
2.	c	12.	a	22.	d	32.	T
3.	c	13.	d	23.	c	33.	F
4.	a	14.	a	24.	a	34.	F
5.	a	15.	b	25.	b	35.	T
6.	c	16.	a	26.	T	36.	F
7.	b	17.	b	27.	F	37.	T
8.	b	18.	b	28.	F	38.	T
9.	a	19.	b	29.	T	39.	T
10.	a	20.	d	30.	T	40.	T

Chapter 31 Lipid Metabolism

We have emphasized in chapter 31 that fat metabolism cannot be divorced from the metabolism of carbohydrates. The interaction of these processes is concisely summarized in figure 31.2. What may not have been quite so obvious is the similarity between the chemistry of the fatty acid cycle and that of the Krebs cycle. And since metabolic pathways may strike students as endless collections of unrelated reactions, we're going to emphasize patterns and relationships whenever we can.

Let's recall some significant features of the Krebs cycle. During the first half of the Krebs cycle, a molecule which will release two carbon dioxide units is synthesized. In the latter part of the Krebs cycle, the remaining molecule is converted back to the original starting material. Here's a summary of these latter reactions.

Krebs cycle (steps 8, 9 and 10)

$$HOOC\text{-}CH_2\text{-}CH_2\text{-}C(=O)\text{-}OH \xrightarrow{FAD \to FADH_2} HOOC\text{-}CH=CH\text{-}C(=O)\text{-}OH \xrightarrow{H_2O} HOOC\text{-}CH(OH)\text{-}CH_2\text{-}C(=O)\text{-}OH \xrightarrow{NAD^+ \to NADH + H^+} HOOC\text{-}C(=O)\text{-}CH_2\text{-}C(=O)\text{-}OH$$

Now look at reactions of the fatty acid cycle.

Fatty acid cycle (excluding final step)

$$R\text{-}CH_2\text{-}CH_2\text{-}C(=O)\text{-}SCoA \xrightarrow{FAD \to FADH_2} R\text{-}CH=CH\text{-}C(=O)\text{-}SCoA \xrightarrow{H_2O} R\text{-}CH(OH)\text{-}CH_2\text{-}C(=O)\text{-}SCoA \xrightarrow{NAD^+ \to NADH + H^+} R\text{-}C(=O)\text{-}CH_2\text{-}C(=O)\text{-}SCoA$$

Except that the fatty acid reacts in the form of its thio ester, the reaction sequence is the same. Two saturated carbons are unsaturated, then oxygen is attached and finally the oxygen is converted to a carbonyl group. In both cases, it is that carbonyl group which reacts in the next step. In the Krebs cycle, the carbonyl group adds acetyl coenzyme A. In the fatty acid cycle, the carbonyl group releases acetyl coenzyme A.

Krebs cycle: $HOOC\text{-}C(=O)\text{-}CH_2\text{-}C(=O)\text{-}OH + H_2O + CH_3\text{-}C(=O)\text{-}SCoA \longrightarrow HOOC\text{-}C(OH)(CH_2\text{-}C(=O)\text{-}OH)\text{-}CH_2\text{-}C(=O)\text{-}OH + CoASH$

Fatty acid cycle: $R\text{-}C(=O)\text{-}CH_2\text{-}C(=O)\text{-}SCoA + CoA\text{-}S\text{-}H \longrightarrow R\text{-}C(=O)\text{-}S\text{-}CoA + CH_3\text{-}C(=O)\text{-}SCoA$

Just as the Krebs cycle yields reduced species (FADH$_2$ and NADH) for the respiratory chain, so does the fatty acid cycle. So both of these processes generate ATP indirectly through oxidative phosphorylation. (Remember that the fatty acid cycle not only supplies FADH$_2$ and NADH formed in every turn of the cycle, but also produces acetyl coenzyme A. That acetal coenzyme A feeds into the Krebs cycle where it generates more FADH$_2$ and NADH.)

The biosynthesis of fatty acids (section 31.4) occurs through either a direct reversal of the fatty acid cycle or through a near reversal of the fatty acid cycle. By "near" we mean that, except for the initial step in each turn of the reverse cycle, the sequence of reactions is essentially the same.

$$\text{-}C(=O)\text{-}CH_2\text{-}C(=O)\text{-} \longrightarrow \text{-}CH(OH)\text{-}CH_2\text{-}C(=O)\text{-} \longrightarrow \text{-}CH=CH\text{-}C(=O)\text{-} \longrightarrow \text{-}CH_2\text{-}CH_2\text{-}C(=O)\text{-}$$

start with the carbonyl   end with saturated carbon

The differences between the direct reversal and near reversal sequences are these. In the near reversal (the main synthetic pathway) the growing fatty acid molecule reacts as an enzyme complex (rather than as a coenzyme A ester). Also, whereas each two-carbon unit is added in the form of acetyl coenzyme A in the direct reversal, it is malonyl coenzyme A which adds (with the loss of carbon dioxide) in the near reversal. And finally in the near reversal the reducing agent is NADPH; in the direct reversal NADH and FADH$_2$ are used.

The hydrolysis of fats (triglycerides) and the biosynthesis of fats were also discussed in the chapter. These reaction sequences are discussed in sections 31.3 and 31.5 respectively. Be sure to look over this material. The biosynthesis of phospholipids can be reviewed by referring to figure 31.8, which summarizes the process.

Remember that phosphate, CDP and coenzyme A, when attached to a molecule, serve to activate that molecule. In this sense, these groups play similar roles.

The rest of the material in this chapter is reviewed in the problems at the end of the chapter.

Self-Test

1. The oxidation of 1 g of fat to carbon dioxide yields
   a. one Calorie   b. four kilocalories   c. nine kilocalories
2. The products obtained from the digestion of fats do not include
   a. glycerol   b. fatty acids   c. monoglycerides   d. diglycerides   e. triglycerides
   f. soaps
3. Which is not true of fats?
   a. They supply more energy per gram than carbohydrates.
   b. They can be stored in greater amounts than carbohydrates.
   c. They can be mobilized more quickly than carbohydrates.
   d. All of the above statements are true.
4. The lymphatic system transports fats
   a. from the small intestine to the blood
   b. from the small intestine to the cells
   c. from the blood to the cells
5. Adipose tissue is not
   a. a storage depot for fats
   b. a protective cushion around vital organs
   c. insulation against loss of body heat
   d. a special kind of metabolically inactive connective tissue
   e. All of the above statements are correct.
6. How is fat stored in the adipose tissue?
   a. as droplets of fat within cells
   b. as bilayers forming cell membranes
   c. complexed with protein molecules
7. Which energy reserve is used first?
   a. depot fat   b. fat stored in the liver   c. glycogen
8. Fatty acids which enter the fatty acid cycle are first activated by their conversion to
   a. coenzyme A thioesters   b. enzyme complexes   c. phosphate esters
9. Which is the product of beta-oxidation?
   a. R-CH=CH-C(=O)-SCoA   b. R-CH(OH)-CH$_2$-C(=O)-SCoA   c. R-C(=O)-CH$_2$-C(=O)-SCoA
10. Which oxidizing agent is employed in the fatty acid cycle?
    a. FADH$_2$ b. NAD$^+$ c. NADP$^+$ d. O$_2$
11. What is the end product obtained from the fatty acid in the fatty acid cycle?
    a. acetyl coenzyme A   b. dihydroxyacetone phosphate   c. malonyl coenzyme A
12. If 5 FADH$_2$ and 6 NADH molecules enter the respiratory chain, how many ATP molecules are produced?
    a. 11   b. 22   c. 27   d. 28   e. 33
13. What is the approximate efficiency of the conversion of energy stored in fatty acid molecules to that stored in ATP molecules?
    a. 10%   b. 50%   c. 90%

14. Which is not a path entered by acetyl coenzyme A?
    a. synthesis of steroids
    b. Krebs cycle
    c. fatty acid synthesis
    d. formation of ketone bodies
    e. glycogen synthesis
    f. Acetyl coenzyme A follows all of the above routes.
15. Which category of foodstuff cannot supply acetyl coenzyme A for the biosynthesis of fatty acids?
    a. carbohydrates   b. fats   c. proteins   d. All 3 types supply acetyl CoA.
16. Before fat biosynthesis begins, glycerol must be activated in the form of
    a. its phosphate ester
    b. its coenzyme A thioester
    c. UDP-glycerol
    d. CDP-glycerol
17. For fat biosynthesis, fatty acids are activated in the form of
    a. their phosphate esters
    b. their coenzyme A thioesters
    c. UDP-acid
    d. CDP-acid
18. In fat biosynthesis, the enzyme phosphatase removes the phosphate group from
    a. glycerol phosphate
    b. dihydroxyacetone phosphate
    c. fatty acid phosphate
    d. phosphatidic acid
19. In phospholipid biosynthesis, the cytidine nucleotide
    a. catalyzes the conversion of ethanolamine to choline
    b. acts as a carrier molecule or activator
    c. picks up a phosphate unit from ATP
20. The ketone bodies do not include
    a. acetoacetic acid   b. acetone   c. β-hydroxybutyric acid   d. pyruvic acid
21. High concentrations of ketone bodies in the blood
    a. cause diabetes mellitus
    b. are a symptom of starvation
    c. increase the pH of the blood
    d. All of the above are correct.
22. Which compound does not serve as an intermediate in the biosynthesis of steroids?
    a. acetyl CoA   b. cholesterol   c. estrone   d. squalene
23. Tay-Sachs disease results from an inability to metabolize
    a. glycerides   b. glycogen   c. glycolipids   d. cholesterol
24. As starvation progresses, the following change occurs.
    a. Acidosis progresses to ketosis.
    b. The brain switches from fat metabolism to carbohydrate metabolism.
    c. A person with a placid nature becomes increasingly violent.
    d. None of the above occurs as the condition progresses.
25. In a person suffering from diabetes mellitus which of the following does not lead to an increase in the production of ketone bodies.
    a. conversion of body tissues to fat metabolism
    b. reliance on gluconeogenesis for glucose required by cells
    c. breakdown of insulin to fatty acids
    d. All of the above lead to increased production of ketone bodies.

True or False

T F 26. Fats are transported in the blood primarily as lipoproteins.
T F 27. Blood lipid levels are not affected by normal body processes and remain relatively constant under most conditions.
T F 28. The glycerol obtained from the hydrolysis of fats is fed into the Embden-Meyerhof pathway as dihydroxyacetone phosphate.
T F 29. Condensation of malonyl coenzyme A with a growing fatty acid chain produces fatty acids with an odd number of carbon atoms.
T F 30. In phospholipid biosynthesis in humans, the cytidine nucleotide usually activates the diglyceride molecule.

T F 31. The ketone bodies are excreted in the urine of healthy individuals.
T F 32. Air hunger accompanies acidosis because the ability of the blood to transport oxygen decreases with decreasing pH.
T F 33. Protein metabolism is high in the early and in the late stages of starvation.
T F 34. Most obesity is due to glandular malfunction.
T F 35. The lipid-storage diseases like Niemann-Pick disease produce arteriosclerosis.

ANSWERS:

1.	c	15.	d	29.	F
2.	e	16.	a	30.	F
3.	c	17.	b	31.	T
4.	a	18.	d	32.	T
5.	d	19.	b	33.	T
6.	a	20.	d	34.	F
7.	c	21.	b	35.	F
8.	a	22.	c		
9.	c	23.	c		
10.	b	24.	d		
11.	a	25.	c		
12.	d	26.	T		
13.	b	27.	F		
14.	e	28.	T		

Chapter 32 Protein Metabolism

The study of the metabolism and biosynthesis of amino acids is both easier and more difficult than the study of comparable processes involving carbohydrates and fats. It is easier because most of the emphasis can be placed on two key reversible reactions, both of which involve the fate of the amine group--oxidative deamination and transamination. Figures 32.2 and 32.4 summarize these processes. Some of your time would be well spent in studying these figures.

The fact that amino acid metabolism neatly blends into carbohydrate metabolism also makes the former a little easier to understand (unless, of course, you managed to survive chapter 30 without really understanding carbohydrate metabolism; if that's the case, as the old saying goes, the chickens have come home to roost). Let's assume the citric acid cycle is not a mystery to you. If that's so, many of the ketoacids which result when amine groups are removed from amino acids are already quite familiar to you.

What makes this topic more difficult than the analogous material covered in chapters 30 and 31 is the fact that the variety of amino acids is considerably greater than the variety of monosaccharides or fatty acids. Thus, all monosaccharides isomerize to common intermediates and then follow identical metabolic pathways. Fatty acids cycle through the same spiral pathways, varying only in the number of turns they take. But there is no common intermediate nor common metabolic pathway for the complete degradation of amino acids. As figure 32.2 indicates, they do all ultimately end up in the citric acid cycle. However, the same figure also indicates that their points of entry are quite varied. We've gotten around this difficulty by simply acknowledging it and not bothering to specify all of the details.

One metabolic pathway which we did cover in detail is the urea cycle, which describes the fate of most of the nitrogen removed from amino acids. The following diagram summarizes the urea cycle by indicating the source of the atoms incorporated in the urea product.

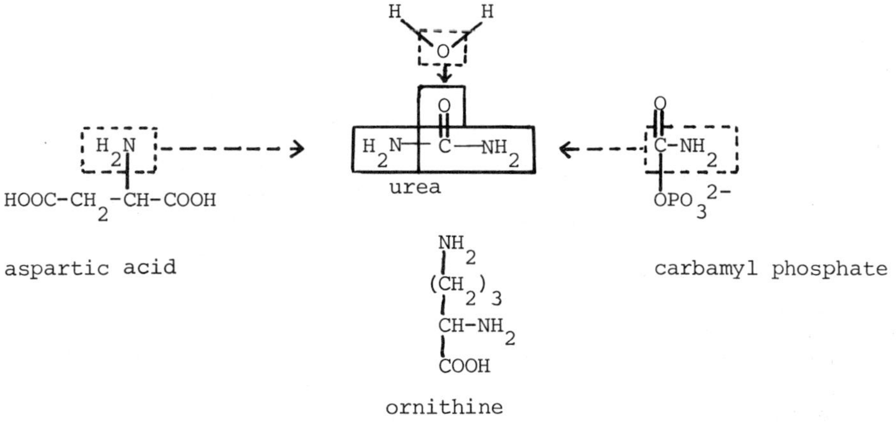

Remember that the nitrogen supplied by aspartic acid may have come from any amino acid by way of transamination reactions. Any amino acid can supply the nitrogen transferred from carbamyl phosphate, too, through a combination of transamination and oxidative deamination. Finally, notice that nothing originates in the ornithine molecule. This compound simply acts as a carrier, pedalling around the cycle, the various ingredients on its back, until the urea is ready to go off on its own.

Figures 32.7 and 32.8 summarize the metabolic fate of purine and pyrimidine bases and can be used to review that material.

As usual, much of the descriptive material in the chapter is reviewed in the problems at the end of the chapter.

Self-Test

1. The biosynthesis of muscle protein from amino acids is classified as
   a. anabolism   b. catabolism   c. digestion   d. transamination
2. The essential amino acids
   a. can be synthesized in the body if nonessential amino acids are supplied in the diet
   b. are not present in the amino acid pool
   c. are the limiting factor in determining the extent of protein biosynthesis
   d. All of the above are correct.
3. Amino acids are stored
   a. with glycogen in liver and muscle
   b. in depots analogous to the fat storage areas
   c. in the nuclei of cells
   d. There are no storage facilities for amino acids in the body.
4. The amino acid pool can be supplied by
   a. dietary amino acids
   b. the breakdown of tissue protein
   c. the biosynthesis of nonessential amino acids
   d. All of the above processes contribute to the population of the amino acid pool.
5. The half-life of collagen molecules in the body is
   a. relatively long
   b. relatively brief
   c. only significant if the collagen is radioactive
6. Protein biosynthesis
   a. occurs only after prolonged starvation has severely depleted body protein
   b. is controlled by the genetic code
   c. occurs primarily in the small intestine
   d. All of the above are correct.
7. Growing children are in a state of
   a. nitrogen balance
   b. positive nitrogen balance
   c. negative nitrogen balance
8. Proteins in the diet can be
   a. used for the production of energy
   b. used in the replacement of tissue protein
   c. converted to carbohydrates
   d. converted to fats
   e. All of the above are true.
9. Which amino acid is most often metabolized via oxidative deamination?
   a. glutamic acid   b. α-ketoglutaric acid   c. ornithine   d. aspartic acid
10. In amino acid metabolism, which compound most commonly serves as the amine group acceptor during transamination reactions?
    a. glutamic acid   b. α-ketoglutaric acid   c. ornithine   d. aspartic acid
11. Which amino acid is converted to oxaloacetic acid through transamination?
    a. glutamic acid   b. α-ketoglutaric acid   c. ornithine   d. aspartic acid
12. The body cannot synthesize essential amino acids because the corresponding ketoacids
    a. will not accept the transfer of an amino group
    b. cannot be synthesized by the body
    c. decarboxylate too quickly
13. Which compound accumulates in the urine of infants suffering from PKU?
    a. phenylalanine   b. tyrosine   c. phenylpyruvic acid   d. melanin
14. GOT is the enzyme which catalyzes the
    a. conversion of phenylalanine to tyrosine
    b. conversion of phenylalanine to phenylpyruvic acid
    c. the transfer of phosphate from creatine phosphate to ADP
    d. the transfer of an amine group from glutamic acid to oxaloacetic acid
15. The term Krebs cycle is not applied to
    a. the citric acid cycle
    b. the urea cycle
    c. the fatty acid cycle
16. The urea formed in the urea cycle does not incorporate atoms contributed by
    a. aspartic acid   b. carbamyl phosphate
    c. ornithine   d. water

17. In addition to urea, which compound is released as a byproduct of the urea cycle?
    a. arginine   b. aspartic acid   c. citrulline   d. fumaric acid
18. Which products of the digestion of nucleic acids are absorbed from the intestine into the blood?
    a. nucleoproteins
    b. nucleotides
    c. nucleosides
    d. pentose sugars and purine and pyrimidine bases
19. Which compounds are converted to uric acid?
    a. adenine and guanine
    b. adenine and thymine
    c. guanine and cytosine
    d. uracil and thymine
20. Which is <u>not</u> an endproduct of the metabolism of purines?
    a. phenylalanine
    b. β-alanine
    c. β-aminoisobutyric acid

True or False

T F 21. The amino acid pool is depleted to obtain material for the synthesis of nitrogen-containing compounds like heme.
T F 22. The turnover rate for proteins of the blood plasma is low.
T F 23. After proper chemical modification, all amino acids can contribute compounds to the tricarboxylic acid cycle.
T F 24. Keto acids formed during the conversion of amino acids to carbohydrates can contribute to the condition known as acidosis.
T F 25. It is possible for the composition of the amino acid pool to be adjusted to fit current needs of the body.
T F 26. All nonessential amino acids are made by transamination reactions.
T F 27. Infants suffering from PKU lack the normal pigmentation in skin, hair and eyes.
T F 28. In human beings, nitrogen is excreted from the body primarily as uric acid.
T F 29. Nucleic acids are not required in the diet of human beings.
T F 30. The precipitation of salts of uracil is characteristics of gout.

ANSWERS:
1. a   11. d   21. T
2. c   12. b   22. F
3. d   13. c   23. T
4. d   14. d   24. T
5. a   15. c   25. T
6. b   16. c   26. F
7. b   17. d   27. F
8. e   18. c   28. F
9. a   19. a   29. T
10. b   20. a   30. F

Part II Answers to Problems in the Text

Chapter 1

1. See section 1.3.
   What distinguishes science from other disciplines is its approach to understanding the world. This scientific method is usually subdivided into three phases:
   (1) gathering data; making observations; taking measurements
   (2) sorting out the relevant facts and establishing a pattern; formulating an hypothesis; developing a model
   (3) testing the hypothesis; devising experiments to determine the predictive value of the model
2. Scientific models are conceptualizations of processes or objects which are not accessible to our senses. We shall soon be speaking of atoms as if they were tiny hard spheres or fuzzy clouds or minute solar systems, although atoms are none of these things. Nonetheless, these models will enable us to predict how real atoms will act under certain circumstances. Model planes or boats, in contrast, are representations of something easily observed. Such models are ordinarily judged on the basis of how closely they resemble the objects they represent.
3. a. 20 °C; -3.9 °C; -23 °C
   b. -24 °F; 414 °F
   c. 310 K; 173 K
4. 50 000; 0.25; 23; 92
5. 10; 0.2; 142; 679
6. 1; 15
7. 1500 cal or 1.5 kcal
8. 140 000 cal or 140 kcal
9. 1.14 g/cm^3
10. 1.6 g/ml
11. 680 g
12. 1.02
13. Section 1.4
    gases--have mass and occupy space, but their volume and shape are determined by their container; can be readily compressed
    liquids--have mass and occupy space; have a relatively fixed volume, but flow readily and adopt the shape of their container
    solids--have mass and occupy space; have relatively fixed volume and shape; are frequently crystalline
14. Kinetic energy is the energy a system contains because of its motion. Potential energy is the energy a system contains because of its position. See section 1.4.
15. Mass is a measure of the amount of matter in an object. Weight is a measure of the gravitational force of attraction between some object and the earth (usually). Section 1.4.
16. Two like-charged particles (both positively charged or both negatively charged) repel one another. Two particles with unlike charges (one positively charged and one negatively charged) attract one another. Section 1.4.
17. --
18. One's weight on the moon is about one-sixth one's weight on earth. (The moon is a smaller body than the earth and the force of gravity at its surface is smaller than at the surface of the earth.) One's mass would be the same on earth and moon.

Chapter 2

1. Most Greeks simply could not envision a piece of matter so minute it could not be divided into still smaller pieces. The Greeks were particularly attracted to the ideal of the most perfect abstractions, and infinite divisibility was such an abstraction.
2. Democritus had neither the tools to test his theories nor an audience receptive to the evidence of such tests. The ancient Greeks preferred to reason from first principles rather than to draw conclusions from physical experiments.
3. A scientific law is a statement which summarizes experimental data. A governmental law is a regulation or rule established by mutual agreement (at least implied agreement) of the citizenry. Governmental laws may be changed at the will of the legislature or may be disobeyed. Scientific laws, since they simply describe natural phenomena, are not subject to human whim -- they cannot be changed or disobeyed at will.
4. A theory is a model which consistently explains the data summarized in scientific laws.
5. The Greeks regarded elements as basic substances which could not be reduced to even more basic substances and counted only four such elements, earth, air, water and fire. Modern elements are the somewhat more than one hundred substances which are distinguished by the atomic number of their constituent atoms.
6. See section 2.5.
7. See section 2.3 and 2.4.
8. From 180 g of water, 20 g of hydrogen are formed.
   To produce 2 tonnes of hydrogen, 18 tonnes of water are required.
   To produce 1 tonne of hydrogen, 9 tonnes of water are required.
9. a. 30 g of carbon
   b. 300 g of carbon
10. 4 hydrogen atoms for each silicon
11. Van Helmont ignored one other possible source of nutrients for the tree -- the air. Plants do require water, but they also require carbon dioxide, and the latter compound they obtain from the atmosphere.
12. The experiments of Thomson, see section 2.5.
13. Rutherford's experiments, see section 2.8.
14. See section 2.7.
15.

Atom	Electrons	Protons	Neutrons
calcium (Ca)	20	20	20
fluorine (F)	9	9	10
beryllium (Be)	4	4	5
sodium (Na)	11	11	12
argon (Ar)	18	18	22
nitrogen (N)	7	7	7

16. a. He (2p, 2n) 2e
    b. Mg (12p, 12n) 2e 8e 2e
    c. C (6p, 6n) 2e 4e
    d. S (16p, 16n) 2e 8e 6e
    e. O (8p, 8n) 2e 6e
    f. Si (14p, 14n) 2e 8e 4e
17. $2n^2 = 2(4^2) = 32$ electrons
18. The Bohr orbits were paths traveled by electrons and located at definite distances from the nucleus of an atom. Quantum mechanical orbitals are mathematical formulations which express the probability of finding an electron within a given volume of space; they are somewhat indefinite locations at which electrons are most likely to be found.
19. For fluorine and chlorine the outermost electron shells contain 7 electrons; fluorine's outermost electrons are in the second shell, while chlorine's are in the third shell.
    Carbon and silicon both have 4 electrons in their outermost shells; carbon's outermost occupied shell is the second shell, silicon's is the third shell. Fluorine has one more electron in its outermost shell than does oxygen.
20. No. To find an element between atomic number 12 and 13 would mean that the atoms of the element contain a fraction of a proton.

Chapter 3

1. a. isotope--atoms of the same element (same atomic number) with different weights (different mass numbers); atoms with the same number of protons but different numbers of neutrons
   b. deuterium--a hydrogen isotope containing one neutron, i.e., ^2_1H
   c. alpha particle--the helium nucleus, ^4_2He
   d. beta particle--an electron, $^0_{-1}e$ or $^0_{-1}\beta$
   e. gamma ray--electromagnetic radiation emitted in nuclear decay; radiation related to visible light, but of much higher energy
   f. half-life--the period of time in which one-half of the atoms in a sample of a radioactive isotope

129

130

      undergo nuclear decay
g. cosmic ray--charged particles of extraterrestrial origin with very high kinetic energy
h. positron--a particle with the same mass as an electron, but with the opposite charge, $^{0}_{1}e$ or $^{0}_{1}\beta$
i. curie (Ci)--a measure of the activity of a sample of a radioactive isotope; 1 curie = 3.7×10^{10} disintegrations/second
j. roentgen (R)--a measure of exposure; a measure of the ionizing effect of x-rays or gamma rays; the amount of x-rays or gamma rays required to produce ions carrying a total of 2.1 billion units of electrical charge in 1 cm^3 of dry air at 0 °C and normal atmospheric pressure
k. rad, radiation absorbed dose--a measure of exposure; the amount of radiation which will impart to each gram of absorbing tissue 100 ergs of energy
l. rem, roentgen equivalent in man--a measure of the biological damage caused by a dose of radiation; the amount of ionizing radiation which produces the same biological damage as one roentgen of x-rays of specified energy
m. Geiger counter--a device for measuring the level or intensity of ionizing radiation; see section 3.9 for a description of the operation of this instrument
n. scintillation counter--a device for measuring radiation levels which operates by detecting and counting light flashes produced by the interaction of ionizing radiation and a phosphor
o. background radiation--the radiation from natural sources (e.g., cosmic rays or uranium ores) to which we are routinely exposed
p. LD$_{50}$/30 days--the dose of radiation required to kill half the population exposed within thirty days
q. fission--a nuclear reaction in which the nucleus of an atom, upon absorption of a neutron, splits into two pieces of roughly comparable size, e.g.: $^{235}_{92}U + ^{1}_{0}n \rightarrow ^{90}_{38}Sr + ^{143}_{54}Xe + 3\, ^{1}_{0}n$ + energy
r. fusion--a nuclear reaction in which light nuclei are fused to form a more massive nucleus at high temperature, e.g.: $^{2}_{1}H + ^{3}_{1}H \rightarrow ^{4}_{2}He + ^{1}_{0}n$ + energy
s. radioisotope--a radioactive isotope
t. radioactive tracer--a material which incorporates a radioactive isotope and can thus be detected as it proceeds through the course of some process, including biochemical reactions
u. binding energy--also called mass defect; the energy released when separate nuclear particles are combined within a nucleus; the energy equivalent of the differences in mass between the separate particles and the particles combined within the nucleus
v. artificial transmutation--a reaction in which the nucleus of one element is converted to the nucleus of another element through bombardment with subatomic particles

2. a. 82 protons, 124 neutrons
   b. 1 protons, 2 neutrons
   c. 27 protons, 33 neutrons
   d. 90 protons, 143 neutrons
3. $^{24}_{12}Mg + ^{1}_{0}n \rightarrow ^{1}_{1}H + ^{24}_{11}Na$
4. $^{215}_{85}At \rightarrow ^{211}_{83}Bi + ^{4}_{2}He$
5. $^{209}_{82}Pb \rightarrow ^{0}_{-1}e + ^{209}_{83}Bi$
6. after 4.5 seconds (one half-life): 1500 atoms left
   after 9.0 seconds (two half-lives): 750 atoms left
   after 13.5 seconds (three half-lives): 375 atoms left
7. See sections 3.12 and 3.14.
8. See section 3.14.
9. a. use shielding
   b. maintain a distance between the radioactive sample and oneself
   c. minimize time of exposure
10. See section 3.14.
11. Alpha particles are more massive than beta particles and tend to release their energy over a shorter path within tissue. The affected tissue is, therefore, subjected to a more concentrated dose of energy, and the potential for causing irreversible damage is higher.
12. See sections 3.16, 3.17, 3.21 and 3.22.
13. See section 3.20.
14. In fusion, two positively charged nuclei must be brought together. To overcome the repulsive forces between such particles, very high temperatures (at which the two particles have enormous kinetic energy) must be used. Section 3.23.
15. See section 3.19.
16. See section 3.22.
17. --

Chapter 4

1. See sections 4.1 and 4.3.
2. See sections 4.1, 4.3 and 4.7.
3. The phrase suggests that sodium chloride exists as a substance composed of discrete subunits consisting of a single sodium ion and a single chloride ion, which is not the case. Sodium chloride is a salt, and salts exist as crystalline substances whose basic structure involves an extended arrangement of ions in a crystal lattice. See section 4.4.
4. a. LiF   g. CaHPO$_4$
   b. CaI$_2$ h. Mg$_3$(PO$_4$)$_2$
   c. AlBr$_3$ i. KNO$_3$
   d. MgSO$_4$ j. Fe(OH)$_3$
   e. (NH$_4$)$_3$PO$_4$ k. Al$_2$S$_3$
   f. Na$_2$C$_2$O$_4$ l. Cu(CN)$_2$
5. a. sodium bromide
   b. calcium chloride
   c. aluminum oxide
   d. calcium sulfate
   e. sodium hydrogen sulfate (sodium bisulfate)
   f. potassium hydrogen carbonate (potassium bicarbonate)
   g. magnesium acetate
   h. aluminum acetate
   i. ammonium phosphate
   j. ammonium oxalate
   k. potassium nitrite
   l. sodium monohydrogen phosphate
   m. potassium cyanide
   n. aluminum hydroxide
   o. calcium dihydrogen phosphate
   p. sodium carbonate
   q. magnesium hydrogen carbonate (magnesium bicarbonate)
   r. ammonium iodide
   s. calcium hydrogen sulfate (calcium bisulfate)
   t. sodium nitrate
6. Monovalent (univalent): any Group IA or Group VIIA element
   Divalent (bivalent): Group IIA or Group VIA elements
   Trivalent: Group IIIA or Group VA elements
   Tetravalent: Group IVA elements
7. a. H$_2$S = H:S:H   Four electron pairs surrounding sulfur adopt a tetrahedral arrangement. Two pairs of electrons are shared by hydrogen atoms; therefore, the molecule should be bent, resembling water.

   b. SiH$_4$ = H:Si:H   The four pairs of electrons around silicon adopt a tetrahedral arrangement. All four pairs are shared by hydrogen atoms and the molecule is tetrahedral in shape like CH$_4$ (methane).

   c. BeCl$_2$ = :Cl:Be:Cl: There are two electron pairs in beryllium's valence shell. The most stable arrangement is linear (180° apart): Cl—Be—Cl

   d. BF$_3$ = :F:B:F: There are three electron pairs in boron's valence shell. The molecule is planar with the electron pairs arranged at angles of 120°.

8.

	:Ẍ·	:Ÿ·	:Z̈·
a.	VIIA	VIA	VA
b.	:Ẍ:H	:Ÿ:H H	H :Z̈:H H
	linear (like HCl)	bent (like H$_2$O)	pyramidal (like NH$_3$)
c.	:Ẍ:⁻	:Ÿ:²⁻	

9.
a. H-C(H)(H)-O-H
b. H-C(H)(H)-N(H)-H
c. H-N(H)-O-H
d. H-N(H)-N(H)-H
e. F-N-F, F
f. H-C(H)=C(H)-H
g. H-C≡C-H

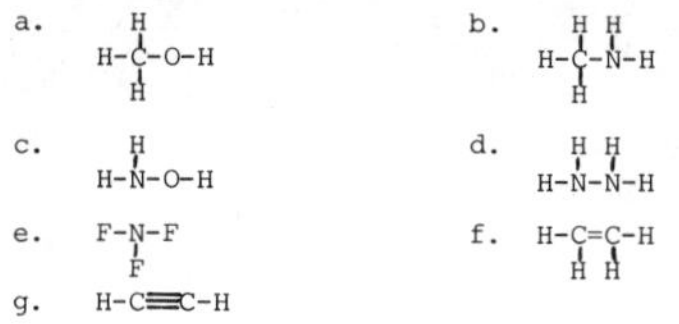

Chapter 5

1.
a. $4 Al + 3 O_2 \longrightarrow 2 Al_2O_3$
b. $2 C + O_2 \longrightarrow 2 CO$
c. $N_2 + O_2 \longrightarrow 2 NO$
d. $2 SO_2 + O_2 \longrightarrow 2 SO_3$
e. $2 NO + O_2 \longrightarrow 2 NO_2$
f. $Zn + 2 HCl \longrightarrow ZnCl_2 + H_2$
g. $Al_2(SO_4)_3 + 6 NaOH \longrightarrow 2 Al(OH)_3 + 3 Na_2SO_4$
h. $Zn(OH)_2 + 2 HNO_3 \longrightarrow Zn(NO_3)_2 + 2 H_2O$
i. $3 NH_4OH + H_3PO_4 \longrightarrow (NH_4)_3PO_4 + 3 H_2O$

2. a. 16 b. 84 c. 80 d. 164 e. 233 f. 352 g. 58 h. 98 i. 123
3. a. 2 moles b. 4 moles c. 0.1 mole d. 0.01 mole e. 0.001 mole f. 0.05 mole g. 0.1 mole h. 10 moles i. 0.03 mole j. 0.0001 mole
4. a. 0.016 g b. 504 g c. 4.0 g d. 246 g e. 699 g f. 14 080 g g. 336 g h. 980 g i. 2.46 g j. 33 g
5. a. 4 moles b. 16 g c. 1.6 g d. 10 ℓ
6. a. 32 g b. 96 kcal
7.
a. energy of activation--the minimum amount of energy which must be supplied to reactants in order to successfully initiate a reaction; the potential energy difference between the reactants' "valley" and the top of the energy "mountain" to be climbed (see Figure 5.8)
b. catalyst--a substance which increases the rate of a reaction without itself being permanently changed; a substance which lowers the activation energy for a given reaction
c. exothermic--a term used to describe a reaction in which heat is one of the products; a reaction in which heat is transferred from the reaction system to the surroundings
d. endothermic--a term used to describe reactions in which heat is absorbed by the reaction system from the surroundings
e. formula weight--the sum of the atomic weights of the atoms in the formula of a substance
f. mole--the amount of a substance containing as many elementary entities as there are atoms in exactly 12 grams of the isotope $^{12}_{6}C$; the amount of a substance which contains 6.02 x 10^{23} of the smallest units of the substance; the amount of a substance equal to one gram formula weight of the substance
g. Avogadro's number--6.02 x 10^{23}; the number of atoms in 12 grams of carbon-12; the number of elementary entities in one mole of a substance
h. reversible reaction--a reaction which can proceed in a forward or reverse direction, from "reactants" to "products" or from "products" to "reactants"; a reaction in which the energy barrier can be approached from either side
i. equilibrium--the state of a reversible reaction in which the rate of the forward reaction exactly equals the rate of the reverse reaction
j. Le Chatelier's principle--a rule applied to reversible reactions which states that, if a stress is applied to a system at equilibrium, the system rearranges in such a way as to minimize the stress

8. See sections 5.10 and 5.12.
9.
a. Increased pressure will shift the equilibrium to the right.
Increased temperature will shift the equilibrium to the left.
b. Increased pressure will shift the equilibrium to the right.
Increased temperature will shift the equilibrium to the left.
c. Increased pressure will shift the equilibrium to the left.
Increased temperature will shift the equilibrium to the right.
10.
a. Equilibrium shifts to right.
b. Equilibrium shifts to right.
c. Equilibrium shifts to right.
d. There is no change in the position of the equilibrium; equilibrium would be reached more rapidly.
11. See section 5.9.
12. A "mechanism" is the step by step process by which reactants are ultimately converted to final products. Section 5.12.

Chapter 6

1.
a. Boyle's law--$PV = k$; $P_1V_1 = P_2V_2$; at constant temperature and mass, the volume of a gas varies inversely with the pressure; see section 6.4
b. Charles' law--$V/T = k$; $V_1/T_1 = V_2/T_2$; at constant pressure and mass, the volume of a gas is directly proportional to the absolute temperature of the gas; see section 6.5
c. Dalton's law--the total pressure of a mixture of gases is equal to the sum of the partial pressures exerted by the constituent gases; $P_{total} = P_1 + P_2 + P_3 + \ldots$; see section 6.8
d. Henry's law--at constant temperature, the solubility of a gas in a liquid is directly proportional to the pressure of the gas at the surface of the liquid; see section 6.9
e. mm Hg--a unit of pressure; 760 mm of Hg is equivalent to one atmosphere or standard pressure
f. psi--pounds per square inch, a unit of pressure; 14.7 psi is equivalent to one atmosphere or standard pressure
g. partial pressure--the pressure exerted by one component of a mixture of gases; the pressure which a constituent of a gas mixture would exert if it alone occupied the same container at the same temperature
h. torr--a unit of pressure; equivalent to a mm Hg
i. diffusion--the process by which one substance (such as a gas) disperses throughout a container, intermixing with another substance if another is present in the space; according to the kinetic-molecular theory, diffusion results from the normal, random motions of the particles of matter
j. the kinetic-molecular theory--the model used to explain the behavior of matter (particularly gases) and which treats matter as collections of individual particles in motion; see section 6.2

2. nitrogen (N_2), oxygen (O_2), argon (Ar), carbon dioxide (CO_2)
Oxygen and carbon dioxide are most important in respiration.
3. The column of air exerting pressure is longer at sea level than at the top of a mountain:

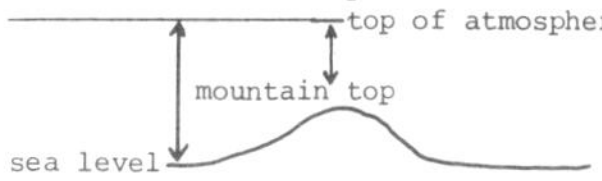

```
                        top of atmosphere
        mountain top
  sea level
```

4. See section 6.3.
5. 200 ft^3

6. 1.25 ℓ
7. Oxygen will flow from flask A to flask B.
   Nitrogen will flow from flask B to flask A.
8. 40 psi
9. 3.3 ℓ
10. 12.7 ℓ
11. 431 ml
12. 22.2 ℓ
13. 2.86 g/ℓ
14. 708 mm Hg
15. 50%
16. In both cases the gases flow with the pressure gradient, i.e., from higher to lower partial pressures. See section 6.10.
17. Carbon monoxide attaches almost irreversibly to hemoglobin and prevents the hemoglobin from transporting oxygen to the cells. See section 6.11.

Chapter 7

1. Gases
   a. highly compressible
   b. individual particles spaced far apart
   c. weak intermolecular forces

   Liquids and Solids
   a. strongly resist compression
   b. individual particles in close contact
   c. compared to gases, strong intermolecular forces; the strength of the intermolecular forces cover a wide range when solids and liquids are compared with one another

2. Solids and liquids occupy relatively fixed volumes, resist compression, experience relatively strong intermolecular forces. Liquids flow readily, while solids tend to maintain their shape. Liquids diffuse more rapidly than solids. See sections 7.6 and 7.8.
3. dipole forces--the intermolecular forces between HCl molecules are dipolar in nature
   hydrogen bonds--water molecules interact through hydrogen bonds
   dispersion forces--chlorine molecules interact through dispersion forces
   interionic forces--salts, like sodium chloride, are held together by interionic forces
   Sections 7.2, 7.3, 7.4 and 7.5.
4. The viscosity of a liquid decreases with increasing temperature. Section 7.6.
5. See section 7.6.
6. Only for b and d would hydrogen bonding be an important intermolecular force in the pure compounds.
7. As the contents of the pressure cooker are heated, water vaporizes. Because the cooker is sealed, the vapor molecules cannot escape to the atmosphere and the vapor pressure builds up in the cooker. The boiling point of water is higher at the higher pressure and, therefore, the temperature inside the pressure cooker can reach higher values than are possible in an open system. See section 7.7.
8. 7200 cal/mole
9. The conversion of a liquid to a vapor requires that particles in contact with one another (liquid) be pulled apart until the distances between them are relatively great (vapor). If the intermolecular forces of attraction are great, then the energy required to pull the particles apart (heat of vaporization) must be great. If the forces are weak, then the heat of vaporization should be small.
10. 1 200 000 cal or 1200 kcal
11. calculated value = 7413 cal/mole; experimental value from Table 7.3 = 7300 cal/mole
12. gold
13. 1000 cal
14. 72 250 cal or 72.25 kcal
15. Section 7.10.
    Some distinctive properties of water are:
        high heat of vaporization
        high specific heat
        high boiling point (it is a liquid at room temperature in contrast to other substances of comparable formula weight)
        the solid is less dense than the liquid
16. See section 7.10.

Chapter 8

1. a. $CH_4 + 2 O_2 \longrightarrow CO_2 + 2 H_2O$
   b. $C + O_2 \longrightarrow CO_2$
   c. $S + O_2 \longrightarrow SO_2$
   d. $CS_2 + 3 O_2 \longrightarrow CO_2 + 2 SO_2$
2. a. 0; b. +6; c. -2; d. -1; e. +3; f. +3; g. +4; h. +4; i. +2; j. +6
3. a. (Cl_2)   KBr
   b. Al   (O_2)
   c. (C_2H_4)   H_2
   d. SO_2   (O_2)
   e. Fe   (HCl)
   f. (HNO_3)   SO_2
   g. C_2H_5OH   (MnO_4)
4. Oxidation
   a. gain of oxygen
   b. loss of hydrogen
   c. loss of electrons
   d. increase in oxidation number

   Reduction
   a. loss of oxygen
   b. gain of hydrogen
   c. gain of electrons
   d. decrease in oxidation number
5. Section 8.6.
   O_2, $KMnO_4$, $Na_2Cr_2O_7$, H_2O_2, HNO_3, halogens, NaOCl
6. Section 8.7.
   H_2, photographic "developer", H_2O_2, carbon, sodium thiosulfate
7. See section 8.9.
8. See section 8.6 and 8.7.
   As an oxidizing agent: $H_2O_2 \longrightarrow H_2O + "O"$
   As a reducing agent: $H_2O_2 \longrightarrow O_2 + "2 H"$
9. See section 8.10.

Chapter 9

1. a. solution--an intimate, homogeneous mixture of two or more kinds of particles (atoms, molecules or ions)
   b. solvent--the substance which dissolves another substance to form a solution; usually present in larger amount
   c. solute--the substance which is dissolved in another substance to form a solution; usually present in smaller amount
   d. miscible--a term describing substances (usually liquids) which form solutions when mixed in any proportions
   e. immiscible--a term describing substances which are mutually insoluble
   f. saturated--a term describing a solution which contains the maximum amount of solute possible under equilibrium conditions at a given temperature and pressure
   g. unsaturated--a term used to describe a solution which contains less than the maximum amount of solute possible under equilibrium conditions at a given temperature and pressure
   h. supersaturated--a term used to describe a solution in which more solute is dissolved than the maximum possible under equilibrium conditions at the same temperature and pressure
   i. soluble--a term used to describe a substance which will dissolve to an appreciable extent in a given solvent
   j. insoluble--a term used to describe a substance which does not dissolve to an appreciable extent in a given solvent
   k. percent by volume--the concentration of a solution expressed as parts of solute per 100 parts of solution with amounts of both solute and solution expressed in volume units; ml solute/100 ml solution
   l. percent by weight--the concentration of a solution expressed as parts of solute per 100 parts of solution with amounts of both solute and solution expressed in units of weight;

133

    m. molarity--the concentration of a solution expressed in moles of solute per litre of solution
    n. dilute--a term used to describe a solution in which the amount of solute is only a small fraction of that in a saturated solution at the same conditions of a temperature and pressure
    o. concentrated--a term used to describe a solution in which the amount of solute is comparatively close to that in a saturated solution at the same conditions of temperature and pressure
    p. milligram percent--a unit of concentration; mg of solute per 100 ml (or 1 dl) of solution
    q. ppm--parts per million, a unit of concentration used for very dilute solutions
    r. ppb--parts per billion, a unit of concentration used for very dilute solutions
2. See sections 9.3 and 9.4 for explanations.
   a, c, d, g, h, i, j, k. soluble
   b, e, f, l, m, n. insoluble
3.

Concentration of solute (g/ℓ)	Molarity (M)	Formula weight of solute
98	1.0	98
32	0.5	64
0.74	0.01	74
2.6	0.1	26
2	0.025	80
120	3	40
17	0.25	68

4. Some properties of true solutions:
   a. The mixture of particles is homogeneous and intimate.
   b. The particles are atoms, molecules or ions.
   c. The solute does not settle on standing.
   d. The solute cannot be filtered from the solution.
   e. A solution is clear and transparent (but not necessarily colorless).
   f. A beam of light passing through the solution is not visible in the solution.
5. No. Nonpolar solvents are unable to interact sufficiently with ionic compounds to overcome the interionic forces in the crystal lattice; the energy released by solvation is less than that required for breaking apart the crystal lattice. Sections 9.3 and 9.4.
6. Water is a polar liquid, capable of hydrogen bonding; motor oil is nonpolar. The rule "like dissolves like" suggests that the interaction between water and motor oil is not favored, and the motor oil should not dissolve in water. Benzene is a nonpolar liquid, as is motor oil. According to the rule "like dissolves like", the motor oil should dissolve in the benzene.
7. See section 9.6.
8. $-1.86^\circ C$; $100.51^\circ C$; $101.02^\circ C$
9. 6 moles; 0.6 mole; 0.006 mole or 6 millimoles
10. 1 g
11. a. Dissolve 5 g of NaOH in 95 ml of H_2O.
   b. Dissolve 117 g of NaCl in water to give exactly 1 ℓ of solution as measured in a volumetric flask.
   c. Dissolve 45 g of $C_6H_{12}O_6$ in water, using a volumetric flask, to give exactly 500 ml of solution.
   d. Dissolve 100 g of $NaHCO_3$ in 900 ml of water.
   e. Dissolve 4.6 g of NaCl in about 495 ml of water.
12. See Table 9.2.
13. a. osmosis--the diffusion of water through a semipermeable membrane to equalize concentrations of solutions separated by the membrane
   b. semipermeable membrane--a membrane or porous sheet which permits some types of molecules to pass through but prevents the passage of others; cellophane is an example, cell membranes are another
   c. dialyzing membrane--a membrane which permits the passage of small molecules and ions, but prevents the passage of large molecules or colloidal particles (like proteins)
   d. hypotonic solution--a solution whose osmotic pressure is lower than that of a reference solution
   e. hypertonic solution--a solution whose osmotic pressure is higher than that of a reference solution
   f. isotonic solution--a solution whose osmotic pressure is identical to that of a reference solution
   g. plasmolysis--the rupture of a cell which occurs when the cell is placed in a hypotonic solution
   h. hemolysis--the rupture of a red blood cell which occurs when the cell is placed in a hypotonic solution
   i. dialysis--the separation of smaller molecules and ions from larger molecules and colloidal particles by selective diffusion through a dialyzing (semipermeable) membrane
   j. crenation--the shriveling of a cell which occurs when the cell is placed in a hypertonic solution
   k. BOD--biochemical oxygen demand, the amount of oxygen taken up by a sample for the oxidation of organic constituents by microorganisms
   l. hydrate--a crystalline compound to which a fixed number of water molecules are bound
   m. anhydrous compound--a compound from which bound water molecules have been driven off; thus, compounds without bound water molecules or water-free compounds
   n. hygroscopic compound--a compound which can pick up and bind water molecules from the atmosphere
   o. efflorescent compound--a compound which will lose bound water molecules to the air on standing in a dry atmosphere
   p. deliquescent compound--a compound which absorbs so much water from the air that it gradually dissolves in the water which accumulates
14. in osmosis: water molecules pass through the membrane
   in dialysis: small ions and molecules (including water) pass through the membrane; large molecules and colloidal particles are stopped by the membrane
15. The dilution of 1:20 is more concentrated.

Chapter 10

1. See section 10.1.
2. the hydrogen ion or proton, H^+
3. H_3O^+; The two (proton and hydronium ion) are frequently used interchangeably. The hydronium ion is the hydrogen ion bonded to a water molecule; it is an attempt to approximate more closely the species as it actually exists in aqueous solutions.
4. From table 10.1:
   monoprotic acids: HNO_3, HCl, CH_3COOH, HCN
   diprotic acids: H_2SO_4 (another would be H_2CO_3)
   triprotic: H_3PO_4, H_3BO_3
5. See table 10.1.
6. hydrobromic acid
7. nitrous acid
8. oxalic acid
9. metasilicic acid = H_2SiO_3; metasilicate ion = SiO_3^{2-}; sodium metasilicate = Na_2SiO_3
10. See section 10.4.
11. hydroxide ion, OH^-
12. strong bases: NaOH, KOH
   weak bases: $Ca(OH)_2$, $Mg(OH)_2$, $NH_3(aq)$
13. Section 10.4.
Very little magnesium hydroxide actually dissolves in aqueous solution (body fluids). Thus, the concentration of hydroxide ion in solution is very low, low enough not to result in injury to body tissues in contact with the solution.
14. a. $Mg + 2 HCl \longrightarrow MgCl_2 + H_2$
   b. $HCl + LiOH \longrightarrow LiCl + H_2O$
   c. $2 Na + 2 H_2O \longrightarrow 2 NaOH + H_2$
   d. $Al(OH)_3 + 3 HCl \longrightarrow AlCl_3 + 3 H_2O$
   e. $NaHCO_3 + HNO_3 \longrightarrow NaNO_3 + H_2O + CO_2$
   f. $CaCO_3 + 2 HBr \longrightarrow CaBr_2 + H_2O + CO_2$
15. a.

        conjugate pair
   CN^- + HCl ⇌ HCN + Cl^-
   stronger base + stronger acid ⇌ weaker acid + weaker base
        conjugate pair

b.

        conjugate pair
   NH_3 + H_2O ⇌ NH_4^+ + OH^-
   weaker base + weaker acid ⇌ stronger acid + stronger base
        conjugate pair

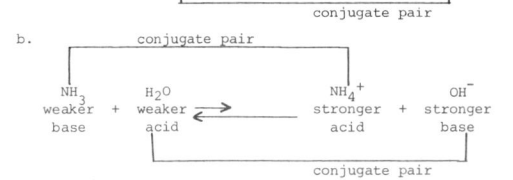

16. hydrolyze (breakdown) protein; dehydrate cells

Chapter 11

1. a. 81, 81   b. 24, 24   c. 34, 17   d. 171, 85.5
   e. 82, 27.3   f. 78, 39   g. 60, 60
2. a. 0.5 $\underline{N}$ b. 1 $\underline{N}$ c. 1 $\underline{N}$ d. 0.04 $\underline{N}$ e. 2 $\underline{N}$
3. 6 $\underline{N}$
4. a. 2 ℓ   b. 1.4 ml
5. a. 0.5 $\underline{N}$ b. 0.25 $\underline{N}$
6. 0.125 $\underline{N}$
7. a. $[H^+] = 10^{-2}$ $\underline{M}$; pH = 2
   b. $[H^+] = 10^{-4}$ $\underline{M}$; pH = 4
   c. $[OH^-] = 10^{-1}$ $\underline{M}$; $[H^+] = 10^{-13}$ $\underline{M}$; pH = 13
   d. $[OH^-] = 10^{-3}$ $\underline{M}$; $[H^+] = 10^{-11}$ $\underline{M}$; pH = 11
   e. $[H^+] = 10^{-5}$ $\underline{M}$; pH = 5
   f. $[OH^-] = 10^{-2}$ $\underline{M}$; $[H^+] = 10^{-12}$ $\underline{M}$; pH = 12
   g. $[H^+] = 10^{-3}$ $\underline{M}$; pH = 3
8. a. basic   b. acidic   c. neutral   d. acidic
9. a. $CH_3COO^- + H_2O \rightleftharpoons CH_3COOH + OH^-$
   b. $CN^- + H_2O \rightleftharpoons HCN + OH^-$
   c. $NH_4^+ + H_2O \rightleftharpoons NH_3 + H_3O^+$
10. a. neutral
    b. basic
    c. can't say; This is a salt of a weak base and a weak acid.
    d. basic
    e. acidic
    f. neutral
11. The products of the titration are water and the salt of a strong base and a weak acid. The solution of such a salt is slightly basic. Section 11.4.
12. See section 11.5.

Chapter 12

1. Definitions
    a. cathode--negatively charged electrode
    b. anode--positively charged electrode
    c. cation--positively charged ion; the type of ion formed when a neutral atom loses electrons
    d. anion--negatively charged ion; the type of ion formed when a neutral atom gains electrons
    e. electrolysis--reaction in which a compound is split into smaller pieces (frequently into its constituent elements) by electricity
    f. strong electrolyte--a compound which will efficiently conduct an electric current when dissolved in water or melted; a compound which is highly ionized in water solutions
    g. weak electrolyte--a compound which will conduct a weak electric current when dissolved in water; a compound which gives a low concentration of ions in water solutions
    h. nonelectrolyte--a compound which fails to conduct an electric current on melting or dissolution
    i. ionization--the formation of ions from covalent compounds upon dissolution, e.g.:
       $HCl + H_2O \longrightarrow H_3O^+ + Cl^-$
    j. ion dissociation--the solvation of an ionic compound; the separation of the constituent ions of an ionic solid which occurs when the solid dissolves
    k. solubility product constant--the product of the concentrations of the ions of a salt in a saturated solution of the salt
2. a. KCl - strong   f. CH_3OH - nonelectrolyte
   b. HNO_3 - strong   g. AgCl - weak
   c. H_2CO_3 - weak   h. CCl_4 - nonelectrolyte
   d. Na_2SO_4 - strong   i. $CaCl_2$ - strong
   e. KOH - strong
3. One mole of NaCl in solution yields approximately two moles of ions, one mole of Na^+ and one mole of Cl^-. One mole of sugar yields one mole of sugar molecules in water. Freezing point depression depends on the total number of particles in solution. (Section 12.3)

There is some association among ions in relatively concentrated solutions. Because of this, some ions do not act completely independently of others. Thus a close association between two ions would result in their behaving as a single particle rather than two independent particles. (Section 12.4)

4. Hydrogen chloride ionizes in aqueous solution to form H_3O^+ and Cl^-. These ions conduct electric current. (Section 12.2)
5. a. K and Br_2 b. Li and Cl_2 c. Al and O_2
6. $Ba(OH)_2 + H_2SO_4 \longrightarrow BaSO_4 + 2 H_2O$
   $Ba^{2+} + 2 OH^- + 2 H^+ + SO_4^{2-} \longrightarrow \underline{BaSO_4} + 2 H_2O$
   Although both reactants are essentially completely ionized in solution, one product is a very slightly soluble salt ($BaSO_4$) and the other is a covalent molecule (H_2O).
   Neither of the products yields significant amounts of ions in solution, and therefore the solution does not conduct electricity effectively.
7. The concentration of Mg^{2+} is 0.001 mole/ℓ or 10^{-3} $\underline{M}$ (from the dissolution of $MgCl_2$). The concentration of CO_3^{2-} is 0.001 mole/ℓ or 10^{-3} $\underline{M}$ (from the dissolution of Na_2CO_3). The ion product of Mg^{2+} and CO_3^{2-} is $(10^{-3})(10^{-3}) = 10^{-6}$. The solubility product constant for $MgCO_3$ is 1×10^{-5}. Since 10^{-6} is smaller than 10^{-5}, the solubility product constant has not been exceeded and no precipitate forms.
8. The object should be the cathode. At the cathode, metal ions pick up electrons and deposit as the metallic element.
9. The solubility product constant for silver acetate is exceeded when the addition of sodium acetate increases the concentration of acetate ions in the solution:
   $Ag^+ + CH_3COO^- \longrightarrow \underline{CH_3COOAg}$
   When acid is added to the solution, some of the dissolved acetate ions react with the protons thus introduced to form the weak acid, acetic acid:
   $CH_3COO^- + H^+ \longrightarrow CH_3COOH$
   This decreases the concentration of acetate ion; the solubility product constant is no longer exceeded and the silver acetate precipitate dissolves:
   $\underline{CH_3COOAg} \longrightarrow CH_3COO^- + Ag^+$
10. $CaCO_3 + 2 HCl \longrightarrow CaCl_2 + H_2O + CO_2$
    insoluble   soluble
11. See section 12.9.

Chapter 13

1. Definitions
    a. inorganic chemistry--the chemistry of the elements other than carbon
    b. organic chemistry--the chemistry of the compounds of carbon
    c. noble gas--the Group O elements; the elements whose outer electron shell contains a stable octet of electrons (except in the case of helium, which has a filled first shell with two electrons)
    d. halogen--any of the Group VIIA elements
    e. alkali metal--any of the Group IA elements
    f. alkaline earth metal--any of the Group IIA elements
2. Electron dot symbols
   a. :Ne: b. :Ö: c. ·F̈:   d. K·   e. Ba·   f. ·N̈·
3. The noble gases are chemically very inert because their outer or valence electron configurations are very stable.
4. Helium is nonflammable, whereas hydrogen carries with it the danger of explosions.
   Helium is less soluble in the blood than nitrogen, therefore the danger of the bends is reduced.
   Helium is a lighter gas than nitrogen and, therefore, diffuses more rapidly.
5. Argon provides an inert atmosphere which preserves the tungsten filament in a light bulb and prolongs its life.
6. The neon, enclosed within glass tubes fitted with electrodes, emits an orange-red light when subjected to an electrical discharge.
7. Electron dot symbols
   a. :F̈:⁻   b. :Ï:⁻   c. :Ö:²⁻   d. :S̈:²⁻
   e. K^+ f. Sr^{2+}

8. a. chlorine - water purification (bacteriocide)
    b. iodine - topical antiseptic
    c. sodium fluoride - pesticide
    d. silver bromide - light sensitive compound used in photographic film
    e. sodium iodide - table salt additive which supplies a necessary nutrient in human diets
    f. ammonia - fertilizer
    g. ammonium nitrate - fertilizer
    h. lithium carbonate - drug used in the treatment of manic depressive psychoses
    i. sodium chloride - table salt
9. See section 13.2.
10. Basic oxides are oxides of metals, e.g., Na_2O or CaO. Acidic oxides are oxides of nonmetals, e.g., NO_2 or SO_3.
11. a. $4\,Li + O_2 \longrightarrow 2\,Li_2O$
    b. $2\,Ca + O_2 \longrightarrow 2\,CaO$
    c. $S + O_2 \longrightarrow SO_2$
    d. $CaO + H_2O \longrightarrow Ca(OH)_2$
    e. $SO_3 + H_2O \longrightarrow H_2SO_4$
    f. $Ca + S \longrightarrow CaS$
12. For a description of photochemical smog see section 13.6. The chemical which initiates the chain of reactions associated with photochemical smog by absorbing sunlight is NO_2.
13. Synergism is the action of two or more substances whose combined effect is greater than the sum of the individual effects.
14. Allotropes are modifications of an element which can exist in more than one form in the same physical state. Carbon exists in three allotropic forms: graphite, diamond and carbon black. Oxygen exists in two allotropic forms: O_2 and O_3, molecular oxygen and ozone.
15. Oxygen atom is O or :Ö:; oxygen molecule is O_2; and ozone is O_3.
    The oxygen atom exists only as a short-lived intermediate of reactions; it is very reactive. Molecular oxygen (O_2) is comparatively stable, but still considered fairly reactive chemically; required for respiration; a colorless, odorless, tasteless gas.
    Ozone is an unstable, toxic, very reactive compound; it is a blue gas with a sweet odor.
16. In the upper atmosphere ozone interacts with sunlight. It absorbs ultraviolet radiation which would otherwise damage living creatures at the earth's surface. At ground level, ozone can cause eye irritation at low concentration and pulmonary edema and hemorrhage at high concentrations.
17. See section 13.4.
18. London smog is characterized by high levels of sulfur oxides in the atmosphere, along with high levels of particulates. It is most prevalent in areas which couple cool, damp weather conditions with the burning of high-sulfur fuel.
    Los Angeles smog is characterized by high levels of nitrogen oxides and other components of automobile exhaust gases. It is most prevalent in areas which couple warm, dry, sunny weather with a high concentration of automobiles.
    The greenhouse effect is the warming effect which results from increased concentrations of carbon dioxide in the atmosphere. Carbon dioxide is transparent to visible light (thus allows sunlight to reach the earth's surface), but opaque in the infrared region (thus traps the longer wavelengths of electromagnetic radiation associated with heat).
19. Nitrogen fixation is the conversion of atmospheric nitrogen (N_2) to more readily utilized forms of nitrogen (such as nitrates or nitrogen oxides). See section 13.5.
20. The industrial fixation of nitrogen to make fertilizer requires large amounts of energy. A petroleum shortage generates an energy shortage which, in turn, can generate a fertilizer shortage.
21. Combustion carried out with insufficient oxygen produces carbon monoxide. See section 13.7.
22. Carbon monoxide bonds essentially irreversibly to hemoglobin, preventing that molecule from carrying oxygen to the cells of the body.
23. Na^+ and K^+
24. Berylliosis is a condition resembling the "black lung" suffered by long-time coal miners which results from the inhalation of beryllium oxide or beryllium metal.
25. the catalyst for photosynthesis, chlorophyll
26. Calcium ion is needed for the formation of bones and teeth, for blood clotting, for the regulation of the heart beat.
27. Hard water is water which contains Ca^{2+}, Mg^{2+}, Fe^{2+} and/or Fe^{3+} ions in relatively high concentrations. These ions combine with soap to produce a curdy precipitate which deposits on clothing, skin, hair, etc. See section 13.9.
28. Three distinguishing characteristics of transitions metals are:
    a. their inner electron levels are being filled
    b. their compounds are usually colored
    c. they exist in several oxidation states
29. See section 13.10.

Chapter 14

1. Three characteristics of the carbon atom:
    a. It can bond to itself to form chains of any length.
    b. It can form strong covalent bonds with many other elements.
    c. It can form many different compounds with the same molecular formula; its compounds exhibit isomerism.
2. Organic Compound   Inorganic Compound
    covalent   ionic
    liquid (low melting)   solid (high melting)
    insoluble in water   soluble in water
    flammable   nonflammable
3. Definitions
    a. hydrocarbons--compounds containing only carbon and hydrogen
    b. alkane--a hydrocarbon containing only single bonds
    c. paraffin--a common name for alkanes
    d. saturated--compounds containing only single bonds; alkanes are an example
    e. unsaturated--compounds containing multiple (double or triple) bonds; alkenes are an example
    f. substituent--a group which is attached to the parent or main chain of an organic molecule
    g. alkene--a hydrocarbon which contains a carbon-carbon double bond
    h. alkyl group--a group or substituent which contains only carbon and hydrogen joined by single bonds; the group which results if a hydrogen is removed from an alkane; CH_3-, CH_3CH_2-, etc.
    i. alkyne--a hydrocarbon containing a carbon-carbon triple bond
    j. homologous series--a series of compounds in which each succeeding member incorporates an additional CH_2 group
    k. isomers--different compounds with the same molecular formula
    l. geometric isomerism--cis-trans isomerism; a form of isomerism resulting from restricted rotation about a double bond; examples are

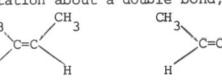

    [cis and trans 2-butene structures]

    m. aromatic compounds--organic compounds related to benzene

4. $CH_3-CH_2-CH_2-CH_2-CH_2-CH_3$ $CH_3-CH-CH_2-CH_2-CH_3$
                                                      |
                                                      CH_3
    hexane   2-methylpentane

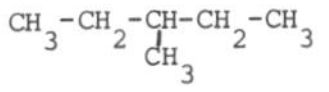

    $CH_3-CH_2-CH-CH_2-CH_3$ $CH_3-CH-CH-CH_3$
              | | |
              CH_3 CH_3 CH_3
    3-methylpentane 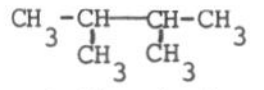   2,3-dimethylbutane

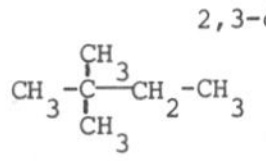

                 CH_3
                  |
    $CH_3-C-CH_2-CH_3$
                  |
                 CH_3
    2,2-dimethylbutane

5. $\quad CH_3$ $\quad CH_3$
           | |
    $CH_2=C-CH_2CH_2CH_3$ $CH_3-C=CH-CH_2-CH_3$
    2-methyl-1-pentene 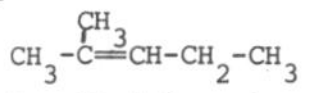   2-methyl-2-pentene

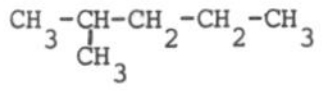

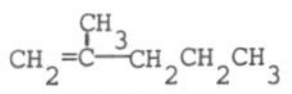

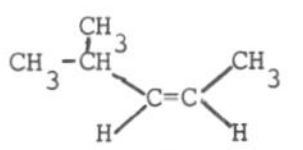

          CH_3 CH_3 CH_3 H
            \   /   \   /
             C=C   C=C
            /   \   /   \
    CH_3-CH H CH_3-CH CH_3
    cis-4-methyl-2-pentene 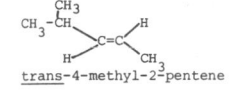   trans-4-methyl-2-pentene

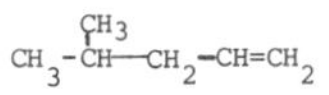

         CH_3
          |
    $CH_3-CH-CH_2-CH=CH_2$
    4-methyl-1-pentene

6. a. CH₃-CH-CH₂-CH₃
        |
        CH₃
    b. CH₃-CH₂-CH-CH-CH₂CH₂CH₃
              | |
              CH₃ CH₃
    c. CH₃ CH₃
       | |
       CH₃-C-CH₂-CH-CH₃
       |
       CH₃
    d. CH₂=CH-CH₃
    e. CH₃-CH=C-CH₃
             |
             CH₃
    f. CH₂=CH-CH₂-CH₂-CH₃
             |
             CH₃ (with extra C-CH₃)
    g. CH₃   CH₂CH₃
          \\C=C/
          /   \\
         H   H
    h. CH₃   H
         \\ /
          C=C
         / \\
        H   CH₂CH₃
    i. H-C≡C-H
    j. HC≡C-CH₂-CH₃
    k. (toluene — benzene with CH₃)
    l. (p-dichlorobenzene — Cl and Cl)
    m. (naphthalene)
    n. (ethylbenzene — CH₃CH₂- on ring, also CH₂CH₃ shown)
    o. (nitro-methyl benzene with O₂N and CH₃ and NO₂)
    p. (isopropylmethylbenzene — H₃C and CH₃ with CH₃ group)
    q. (cyclohexane)
    r. (cyclopentene)

7. a. 3-methylhexane   b. 2-methyl-1-pentene
    c. methylcyclopropane   d. cyclobutene
    e. ethylbenzene   f. cis-3-hexene
    g. 1-pentyne   h. 2,5-dimethyl-2-hexene
    i. 2-butyne   j. 4-isopropylheptane
    k. m-nitrotoluene   l. 1,3,5-trinitrobenzene

8. a. C₃H₈ + 5 O₂ ⟶ 3 CO₂ + 4 H₂O
    b. CH₂=CH-CH₃ + H₂ —Ni→ CH₃-CH₂-CH₃
    c. CH₂=C-CH₃ + Br₂ ⟶ CH₃
                              |
                         CH₂-C-CH₃
                          | |
                          Br Br
    d. CH₂=CH₂ + H₂O —H⁺→ CH₂-CH₂ (or CH₃-CH₂-OH)
                          | |
                          H   OH

9. If alkanes are aspirated into the lungs, they cause "chemical pneumonia" by dissolving fat-like material from the cell membranes of the alveoli. See section 14.7.
10. Cyclopropane is a very effective anesthetic.
11. The lighter (lower boiling) alkanes dissolve natural oils from the skin and cause dermatitis. The heavier liquids (higher boiling) act as emollients and lubricate the skin.
12. Some polycyclic aromatic hydrocarbons such as benzpyrene are carcinogenic, i.e., they can induce cancer.

Chapter 15

1. Aliphatic compounds are those which do not exhibit aromatic properties. (The term aliphatic was originally derived from words which referred to the source of most of these compounds, fats.) The symbol R represents any aliphatic group. Aryl is a general term for an aromatic group; a group with a bond directly to the aromatic ring. The symbol Ar represents any aryl group.

2. CH₂-CH₂-CH₃   CH₃-CH-CH₃
    | |
    Br   Br
    n-propyl bromide   isopropyl bromide
    1-bromopropane   2-bromopropane

3. CH₂CH₂CH₂CH₃   CH₃CHCH₂CH₃
    | |
    Cl   Cl
    n-butyl chloride   sec-butyl chloride
    1-chlorobutane   2-chlorobutane

        CH₃   CH₃
        | |
    CH₂CHCH₃   CH₃CCH₃
    | |
    Cl   Cl
    isobutyl chloride   tert-butyl chloride
    1-chloro-2-methylpropane 2-chloro-2-methylpropane

4. a. CF₃CF₂CF₃   b. CH₂=CH-F
    c. CCl₄   d. CHCl₃
    e. CH₂Cl₂   f. Cl-C=CH-CH₃
                                    |
                                    Cl
    g. (bromo-methyl-bromo-benzene with Br, CH₃, Br)

5. a. 2-bromopentane   b. 3-chloro-2-methylhexane
    c. m-dichlorobenzene   d. 1,1-dichloroethane
    e. 1,1,2-trichloroethane f. 1,2,4-tribromobenzene
    g. 2,4,6-trichloroheptane

6. See section 15.4.
7. Chloroform is nonflammable whereas ether can form explosive mixtures with air. Chloroform is no longer used as an anesthetic because its effective dose is quite close to the lethal dose; it is a toxic compound and a suspected carcinogen.
8. Freons are low molecular weight fluorocarbons, many containing chlorine. Their use as propellants in aerosol spray cans has been called into question. See section 15.6.
9. --
10. --
11. The halogenated hydrocarbons are relatively inert and, therefore, persist long enough to be effective against pests (but also contaminate the environment over long time periods). They are fat-soluble and therefore can be readily picked up and retained by pests to which the compounds are toxic. (The solubility also permits them to be concentrated up the food chain and, therefore, makes them dangerous to species whose elimination is not desirable.) See section 15.8.
12. The present DDT level in humans is about 12 ppm. This level will probably increase for a while as more of the material already in the environment is gradually concentrated up the food chain. If the restrictions on the use of DDT are maintained indefinitely, it is likely that levels will decrease over a long period.
13. 30 grams (about an ounce)
14. See section 15.8.
15. birds and fish
16. Some resistant strains, capable of detoxifying DDT, have developed. See section 15.9.
17. The PCBs and DDT are structurally similar and exhibit similar solubilities and chemical properties.
18. Methoxychlor, Lindane (gamma-benzene hexachloride), Chlordane, Heptachlor, Aldrin, Dieldrin, Endrin

Chapter 16

1. A functional group is a group of atoms within a molecule which confers characteristic chemical and physical properties on a family of organic compounds. The functional group in alcohols is -OH; in alkenes, C=C.

2. and 3.

    CH₂CH₂CH₂CH₂CH₃ (1°)   CH₃
    | |
    OH   CH₃-CH-CH₂CH₃ (1°)
                                   |
                                   OH

    CH₃CH-CH₂CH₂CH₃ (2°)   CH₃
        | |
        OH   CH₃-C-CH₂CH₃ (3°)
                                   |
                                   OH

    CH₃CH₂CHCH₂CH₃ (2°)   CH₃
           | |
           OH   CH₃-CH-CH-CH₃ (2°)
                                      |
                                      OH

         CH₃
         |
    CH₂-C-CH₃ (1°)   CH₃-CH-CH₂-CH₂ (1°)
    | | | |
    OH CH₃   CH₃   OH

4. a. 2-hexanol   b. di-n-propyl ether
    c. ethyl isopropyl ether d. m-chlorophenol (3-chlorophenol)
    e. o-bromophenol (2-bromophenol)   f. isobutyl alcohol (common)
                                           2-methyl-1-propanol (IUPAC)
    g. isopropyl alcohol (common)   h. n-hexyl alcohol (common)
       2-propanol (IUPAC)   1-hexanol (IUPAC)

5. a.   CH₃   b. (I—phenol—OH)
         |
    CH₃-O-C-CH₃
         |
         CH₃
    c. CH₂-CH-CH₃   d. CH₂-CH-CH₂
       | | | | |
       OH OH   OH OH OH
    e. CH₃-CH-CH₂-CH-CH₃   f. CH₃-CH-CH₂-CH₃
          | | |
          OH   CH₃   OH

g. $CH_2-CH_2CH_2CH_3$
   |
   OH

h. $CH_3CH_2CHCH_2CH_2CH_3$
   |
   OH

i. $CH_3-CH-O-CH_3$
   | |
   CH_3 CH_3

j. CH_3
   |
   CH_2-C-CH_3
   | |
   OH CH_3

k. [phenol structure] —OH

6. a. ethanol b. methanol c. 2-propanol
   Methanol (wood alcohol) is most toxic; ethanol (grain alcohol) is least toxic. See section 16.5 and table 16.2.

7. Denatured alcohol is ethyl alcohol to which some material has been added in order to make a mixture which is unfit to drink. Alcohol is denatured to avoid the requirement for paying the luxury tax imposed on drinking alcohol.

8. Distilled spirits. Alcohol from simple fermentation never reaches concentrations greater than 14-18%. This corresponds to a proof of 28-36. To obtain 90 proof liquor the fermentation product must be concentrated by distillation.

9. Methanol's metabolic product, formaldehyde, is more toxic than the acetaldehyde obtained from ethyl alcohol.

10. If you were careless, you'd conclude that methanol was probably safe for human consumption. If you remembered the rule that toxicology data should be interpreted within strict limits, then your conclusion would have been that the data is not sufficient for a conclusion regarding human toxicity.

11. Section 16.7. Ethylene glycol is oxidized to oxalic acid:
    OH OH   O O
    | | [O] || ||
    CH_2-CH_2 ——→ HO-C-C-OH

    The product crystallizes as a calcium salt in the kidneys, where it causes potentially fatal renal damage.
    Propylene glycol is oxidized to pyruvic acid in the body:
    OH OH   O O
    | | [O] || ||
    $CH_3-CH-CH_2$ ——→ $CH_3-C-C-OH$

    Pyruvic acid is a normal intermediate in human metabolism.

12. Ethanol. See section 16.5.

13. a. oxidation   b. dehydration
    c. oxidation   d. hydration
    e. dehydration

14. $CH_3CH_2CH_2CH_2$ ——→ $CH_3CH_2CH_2C-H$ [——→ $CH_3CH_2CH_2C-OH$]
    | || ||
    OH   O   O

    $CH_3CH_2CHCH_3$ ——→ $CH_3CH_2-C-CH_3$
    | ||
    OH   O

    CH_3 CH_3 CH_3
    | | |
    $CH_3-CH-CH_2$ ——→ $CH_3-CH-C-H$ [——→ $CH_3-CH-C-OH$]
    | || ||
    OH   O   O

    CH_3
    |
    CH_3-C-CH_3 ———→ no reaction
    |
    OH

15. $CH_2CH_2CH_3$ $\xrightarrow{H^+}$ $CH_2=CH-CH_3 + H_2O$
    |
    OH

16. 2 $CH_2CH_2CH_3$ $\xrightarrow{H^+}$ $CH_3CH_2CH_2-O-CH_2CH_2CH_3 + H_2O$
       |
       OH

17. Markovnikov's Rule: A rule which predicts the preferred product of the addition of water (and a number of other reagents) to an unsymmetrical alkene. The hydrogen of the water molecule will become attached to the doubly bonded carbon which already has more hydrogens attached. "The rich get richer."
    Example: $CH_2=CH-CH_3$ not $CH_2=CH-CH_3$
             ↑   ↑   ↑   ↑
             H   OH   HO   H

18. a. $CH_2=CHCH_2CH_3$ $\xrightarrow{H^+, H_2O}$ $CH_3-CH-CH_2CH_3$
                                                      |
                                                      OH

    b. [steroid structure] $\xrightarrow{H^+, H_2O}$ [steroid structure with OH]

19. Lowest Boiling ——————→ Highest Boiling
    a. methanol   < ethanol   < 1-propanol
    b. n-butane   < 1-propanol < ethylene glycol
    c. diethyl ether < 1-butanol < propylene glycol

20. Alcohol molecules can hydrogen bond with one another. These strong intermolecular forces should result in a higher boiling point for alcohols relative to aldehydes of comparable molecular weight. The aldehyde molecules cannot hydrogen bond to one another.

21. Increasing solubility ———→
    a. 1-octanol < 1-butanol < methanol
    b. n-pentane < diethyl ether < propylene glycol

Chapter 17

1. $CH_3CH_2CH_2CH_2CH=O$
   valeraldehyde
   pentanal

   CH_3
   |
   $CH_3CHCH_2CH=O$
   β-methylbutyraldehyde
   3-methylbutanal

   CH_3
   |
   $CH_3CH_2CHCH=O$
   α-methylbutyraldehyde
   2-methylbutanal

   CH_3
   |
   $CH_3CCH=O$
   |
   CH_3
   dimethylpropionaldehyde
   dimethylpropanal

2. $CH_3CH_2CH_2\overset{O}{\overset{||}{C}}CH_3$ $CH_3CH_2\overset{O}{\overset{||}{C}}CH_2CH_3$
   methyl n-propyl ketone   diethyl ketone
   2-pentanone   3-pentanone

   $CH_3\overset{O}{\overset{||}{C}}CHCH_3$
                |
                CH_3
   methyl isopropyl ketone
   3-methyl-2-butanone

3. a. benzaldehyde   b. cyclopentanone
   c. ethyl isobutyl ketone   d. propionaldehyde
      5-methyl-3-hexanone   propanal
   e. β,β-dimethylbutyraldehyde
      3,3-dimethylbutanal

4. a. $CH_3\overset{H}{\underset{}{C}}=O + H_2O \longrightarrow CH_3\overset{H}{\underset{OH}{C}}-OH$

   b. $CH_3\overset{H}{\underset{}{C}}=O + CH_3OH \longrightarrow CH_3\overset{H}{\underset{OCH_3}{C}}-OH$

   c. $CH_3\overset{H}{\underset{}{C}}=O + 2\ CH_3OH \xrightarrow{dry\ HCl} CH_3\overset{H}{\underset{OCH_3}{C}}-OCH_3 + H_2O$

   d. $CH_3\overset{H}{\underset{}{C}}=O + H_2N-NH-\text{Ph} \longrightarrow CH_3\overset{H}{\underset{}{C}}=N-NH-\text{Ph} + H_2O$

   e. $2\ CH_3\overset{H}{\underset{}{C}}=O \xrightarrow{NaOH} CH_3-\overset{OH}{\underset{}{CH}}-CH_2-\overset{H}{\underset{}{C}}=O$

5. a. $CH_3\overset{CH_3}{\underset{}{C}}=O + H_2O \longrightarrow CH_3\overset{CH_3}{\underset{OH}{C}}-OH$

   b. $CH_3\overset{CH_3}{\underset{}{C}}=O + CH_3OH \longrightarrow CH_3\overset{CH_3}{\underset{OCH_3}{C}}-OH$

   c. $CH_3\overset{CH_3}{\underset{}{C}}=O + 2\ CH_3OH \xrightarrow{dry\ HCl} CH_3\overset{CH_3}{\underset{OCH_3}{C}}-OCH_3 + H_2O$

   d. $CH_3\overset{CH_3}{\underset{}{C}}=O + H_2N-NH-\text{Ph} \longrightarrow CH_3\overset{CH_3}{\underset{}{C}}=N-NH-\text{Ph} + H_2O$

   e. $2\ CH_3\overset{CH_3}{\underset{}{C}}=O \longrightarrow CH_3-\overset{CH_3}{\underset{OH}{C}}-CH_2-\overset{CH_3}{\underset{}{C}}=O$

6. 2-Pentanone would react with 2,4-dinitrophenylhydrazine; 2-pentanol would not. The addition of the clear orange solution of the reagent would yield an orange precipitate with the ketone, but would remain clear when added to the alcohol.

7. Tollens' reagent. With acetone, no appreciable change would be noted in the clear, colorless Tollens reagent. With the aldehyde, a black precipitate of silver would appear or, if the test tube were very clean, a silver mirror would form on the inner surface of the test tube.

8. $CH_3-\overset{O}{\overset{||}{C}}-CH_3 + OH^- \longrightarrow {}^-CH_2-\overset{O}{\overset{||}{C}}-CH_3 + H_2O$

9. $CH_2=\overset{}{\underset{OH}{C}}-CH_3$

10. | Compound | Reaction to Tollens | Reaction with 2,4-DNP |
    |---|---|---|
    | benzaldehyde | positive | forms derivative |
    | cinnamaldehyde | positive | forms derivative |
    | camphor | negative | forms derivative |

137

138

Compound	Reaction to Tollens	Reaction with 2,4-DNP
irone	negative	forms derivative
vanillin	positive	forms derivative
muscone	negative	forms derivative
diacetyl	negative	forms derivative
cis-3-hexenal	positive	forms derivative
trans,cis-2,6-nonadienal	positive	forms derivative
progesterone	negative	forms derivative
testosterone	negative	forms derivative

11. See section 17.12.

Chapter 18

1. a. formic acid (common)   b. propionic acid
       methanoic acid (IUPAC)   propanoic acid
    c. α,γ-dimethylvaleric acid d. capric acid
       2,4-dimethylpentanoic acid   decanoic acid
    e. glutaric acid
       pentanedioic acid
2. a. sodium benzoate   b. calcium propionate
    c. ammonium acetate   d. zinc butyrate
    e. calcium phthalate
3. Benzamide: C₆H₅C(O)NH₂
    Methyl benzoate: C₆H₅C(O)OCH₃
4. a. CH₃CHCH₂COOH   b. Cl-C₆H₃(Cl)-COOH
          |
          CH₃
    c. CH₃CH₂CH₂CH₂CH₂COOH d. CH₃CHCH₂COOH
                                   |
                                   Cl
    e. OH   f. O
       | ‖
       CH₃CHCH₂COOH   CH₃CH₂OCCH₂CH₃
    g. CH₃CH₂O-C(O)-C(O)-OCH₂CH₃   h. H-C(O)-O⁻ K⁺
    i. CH₃CH₂CH₂CH₂CH₂CH₂O-C(O)CH₃   j. CH₃CH₂C(O)-NH₂
    k. CH₃CH-O-C(O)CH₃   l. CH₃C(O)-NHCH₃
          |
          CH₃
    m. C₆H₅-C(O)-N(CH₃)₂   n. C₆H₅-O-C(O)-CH₃
    o. C₆H₅-C(O)-O-CH₂CH₃ p. C₆H₅-C(O)-NH-C₆H₅

5. a. n-propyl propionate   b. N-isopropylacetamide
    c. isobutyl benzoate   d. phenyl butyrate
    e. benzamide   f. butyranilide
6. lowest boiling ─────────────────► highest boiling
    n-pentane < methyl acetate < n-butyl alcohol < propionic acid
7. a. CH₃CH₂CH₂OH   b. CH₃CH₂CH₂CH₂
                                 | |
                                 OH OH
    c. CH₃OH   d. CH₃CHCH₂CH₂OH
                                 |
                                 CH₃
8. least acidic ───────────────────► most acidic
    toluene < benzyl alcohol < benzoic acid
9. a. CH₃CH₂CH₂CH₂CH₂CH₂CH₂CH₂CH₂COOH + NaOH ⟶
         CH₃CH₂CH₂CH₂CH₂CH₂CH₂CH₂CH₂COO⁻Na⁺ + H₂O

    CH₃CH₂CH₂CH₂CH₂CH₂CH₂CH₂CH₂COOH + NaHCO₃ ⟶
       CH₃CH₂CH₂CH₂CH₂CH₂CH₂CH₂CH₂COO⁻Na⁺ + CO₂ + H₂O

    In both instances, the solid (decanoic acid) would dissolve to give clear solutions of the salt. In the second reaction, bubbles of carbon dioxide gas could be seen escaping from the solution.

    b. C₆H₅-COOH + NaOH ⟶ C₆H₅-COO⁻Na⁺ + H₂O
       C₆H₅-COOH + NaHCO₃ ⟶ C₆H₅-COO⁻Na⁺ + H₂O + CO₂

In both cases, the white solid (benzoic acid) would dissolve to give a clear solution of the salt. In the second reaction, the carbon dioxide gas could be observed as a "fizzing."

c. C₆H₅-CH₂OH + NaOH ⟶ no reaction
   C₆H₅-CH₂OH + NaHCO₃ ⟶ no reaction

In both cases, the colorless liquid (benzyl alcohol) would form a separate layer of liquid when the aqueous solutions were added. See figure 18.2.

10. a. CH₃C(O)-OH and CH₃CH₂OH
    b. CH₃C(O)-O⁻ and CH₃CH₂OH
11. a. C₆H₅-C(O)-OH and CH₃CH₂OH
    b. C₆H₅-C(O)-O⁻ and CH₃CH₂OH
12. a. C₆H₅-C(O)-OH and NH₄⁺
    b. C₆H₅-C(O)-O⁻ and NH₃
13. See section 18.7. Morphine contains two hydroxyl groups (a phenol and an alcohol functional group); these two groups are esterified with acetic acid in heroin.
14. --
15. See section 18.7. Aspirin is acetylsalicylic acid,

    C₆H₄(COOH)(O-C(O)-CH₃)

    Brands may differ in the way in which the aspirin is prepared and purified. The binder which holds the tablets together may vary. The dosage per tablet may differ, and, of course, the cost may differ. All brands must, however, contain the same compound, acetylsalicylic acid.
16. --

Chapter 19

1. | Structure | Name | Classification |
    |---|---|---|
    | CH₃CH₂CH₂CH₂NH₂ | n-butylamine | 1° |
    | CH₃CH₂CHCH₃ with NH₂ | sec-butylamine | 1° |
    | (CH₃)CHCH₂NH₂ | isobutylamine | 1° |
    | (CH₃)₃C-NH₂ | tert-butylamine | 1° |
    | CH₃CH₂CH₂NHCH₃ | methyl-n-propylamine | 2° |
    | CH₃CH(CH₃)-NH-CH₃ | methylisopropylamine | 2° |
    | CH₃CH₂NHCH₂CH₃ | diethylamine | 2° |
    | CH₃CH₂N(CH₃)CH₃ | dimethylethylamine | 3° |

2. A carbocyclic compound contains rings incorporating only carbon atoms. Heterocyclic compounds contain rings which include carbon and elements other than carbon (such as N, O, or S). Examples are:

                carbocyclic   heterocyclic

```
                CH₂—CH₂                        CH₂—CH₂
               /       \                       /       \
              CH₂      CH₂                    CH₂      CH₂
               \       /                       \       /
                 CH₂                             NH
              cyclopentane                    pyrrolidine
```

3. a. CH₃-NH-CH₃ b. CH₃CH₂-CH-CH₂-CH₃
                              |
                              NH₂

   c. [cyclohexane]-NH₂ d. H₂NCH₂CH₂CH₂CH₂CH₂NH₂

   e. [benzene]-NH₂ f. [benzene with Br]-NH₂

   g. [benzene]-N(CH₃)₂ h. CH₂-CH₂
                            | |
                            OH   NH₂

   i. CH₃CH₂-N-CH₂CH₃ j. CH₃CH₂-CH-C-OH
              | | ‖
              CH₃   NH₂ O

   k. [pyridine] l. [purine structure]

   m. [pyrimidine] n. [benzene]-NH₃⁺ Cl⁻

   o. [benzene]-NH₃⁺ Br⁻   [[benzene]-NH₂·HCl]

   p. [CH₃-N(CH₃)₂-CH₃]⁺ Cl⁻

4. Alkaloids are organic nitrogen-containing bases isolated from various plant sources. Examples are caffeine, nicotine and cocaine.
5. a. <u>n</u>-butylamine--Both compounds have comparable molecular weights, but <u>n</u>-butylamine is an associated liquid, i.e., forms strong hydrogen bonds.
   b. <u>n</u>-butyl alcohol--Alcohols have higher boiling points than 1° amines of comparable molecular weight. The oxygen of the alcohol is the more electronegative element and can form stronger hydrogen bonds than the nitrogen of the amine.
   c. <u>n</u>-propylamine--Both compounds have comparable molecular weights, but the 3° amine has no hydrogen attached to nitrogen whereas the 1° amine does. Thus the 1° amine can form hydrogen bonds and has the higher boiling point.
   See section 19.4.
6. See section 19.5. In hydrochloric acid the amine forms an ionic compound, a salt. The salt is more readily dissolved by water than is the neutral amine itself.
7. a. CH₃CHCH₂NH₂   b. CH₃CH₂CH₂NH₂
          |
          CH₃
       isobutylamine   <u>n</u>-propylamine

   c. CH₃CHCH₂CCH₃
          | |
          CH₃ NH₂
       2-amino-4-methylpentane

8. a. [CH₃CH₂-N(H)(H)-CH₂CH₃]⁺ Cl⁻   b. [benzene]-C(=O)-NH₂
   c. [benzene]-C(=O)-NH₂   d. [benzene]-NH-C(=O)-CH₃

   e. CH₃-N-CH₂CH₃
          |
          N=O

9. See section 19.6.
10. a. mixture   b. single compound
    c. single compound   d. single compound
    e. mixture   f. single compound
    g. mixture of two isomers h. single compound
11. Both caffeine and nicotine may be mildly addictive. See section 19.10.
12. Some possibilities: The digestive process may act to detoxify the nicotine before it reaches the blood stream or nicotine may not be efficiently absorbed from the digestive tract into the bloodstream.
13. The minimum lethal dose for a 70 kg person is 350 mg or 0.35 g or about 0.01 ounce. Toxicity studies on animals may not give valid data for humans because the metabolic fate of a substance may be different in humans.
14. Section 19.9. The basic barbiturate structure is:

```
         H    O
         |    ‖
         N----C
        / \  / \
    O=C    C    R
        \ / \ /
         N   R
         |
         H
         (with C=O on right)
```

    The R groups on this basic structure are varied to obtain long- or short-acting drugs, etc.
15. As used in the chapter, synergism is the enhancement of the effect of one drug by another. See section 19.9.
16. See section 19.11.
17. According to table 19.3, cocaine is more toxic. The toxicities of lidocaine and cocaine cannot be compared on the basis of the data given in table 19.3. Both the method of administration and the test animals used in the experiments to determine toxicity were different.
18. See section 19.11. Some examples are halothane, ether, cyclopropane, chloroform or divinyl ether.
19. Curare causes a complete relaxation of muscles, with death caused by respiratory failure. In surgical usage, dose is carefully controlled and the patient can be supported with respiratory therapy. Section 19.11.
20. Tranquilizers are used to control the symptoms of the disease, but are not cures. See section 19.12.
21. Norepinephrine and serotonin. Section 19.7.
22.

[Complex structure with labels: CH₃O— ether, amine, N-H amine, CH₃O-C ester, OCH₃ ether, OCH₃ ether, OCH₃ ether, OCH₃ ether]

Chapter 20

1. a. CH₃CH₂-SH   b. CH₃-SH
   c. CH₃CHCH₃   d. CH₃CH₂CH₂CH₂-S-CH₂CH₂CH₃
         |
         SH
   e. CH₃-S-S-CH₃   f. CH₃CH₂-S(=O)-CH₂CH₃
   g. [benzene]-S(=O)(=O)-[benzene]   h. [benzene]-SO₃H
   i. [benzene]-SO₂-NH₂   j. CH₃CH₂-P-CH₂CH₃
                                   |
                                   CH₂CH₃

k. CH₃CH₂O-P(OH)(=O)-OCH₃
l. CH₃O-P(OCH₃)(=O)-OCH₃
m. HO-P(OH)(=O)-O-P(OH)(=O)-O-P(OH)(=O)-OH

2. a. propanethiol; n-propyl mercaptan
    b. methanesulfonic acid
    c. dimethyl sulfoxide
    d. diethyl sulfone
    e. diethyl disulfide
    f. ethyl methyl sulfide
    g. pyrophosphoric acid
    h. trimethylphosphine
    i. ethyl dihydrogen phosphate
    j. o-chlorobenzenesulfonamide

3. a. CH₃-CH(CH₃)-S-S-CH(CH₃)-CH₃
    b. 2 CH₃-SH
    c. CH₃CH₂-S-CH₂CH₃
    d. CH₃CH₂-S(=O)-CH₂CH₃
    e. C₆H₅-SO₃H
    f. C₆H₅-SO₃⁻ Na⁺
    g. CH₃-S-Hg-S-CH₃

4. Section 20.7. The ABS detergents contain a branched hydrocarbon chain. The LAS detergents contain an unbranched chain. Microorganisms can break down the unbranched compound, but cannot degrade the ABS molecule.

5. Section 20.11. The acetylcholine cycle can be stopped at three points:
   1) by interfering with the synthesis of the messenger molecule acetylcholine - Botulinus works in this way.
   2) by blocking the receptor site and preventing the acetylcholine message from being received - Curare and atropine act in this way.
   3) by preventing the hydrolysis and deactivation of acetylcholine - organophosphorus compounds like Sarin or Tabun act in this way.

6. Section 20.2. BAL chelates mercury ions and prevents them from interacting with enzyme molecules in the body.

7. The nerve poison acts by preventing the hydrolysis (deactivation) of acetylcholine. Therefore, the receptor nerves are being continuously stimulated by acetylcholine and the resulting overstimulation can prove fatal. Atropine acts by blocking the receptor sites. It will counteract the effect of the nerve poisons by simply preventing the receptor nerves from responding to the continuous stimulation of the acetylcholine. The situation here is not unlike the use of ear plugs to block out unwanted noise. The noise is still there, you just don't hear it any more. The atropine acts like an "ear plug" for the nerves. The acetylcholine (noise) is still there; the nerves just won't respond to its presence any more.

Chapter 21

1. a. macromolecule--literally, a large molecule; a term frequently used to describe high molecular weight polymers and molecules of similar size
    b. polymer--large molecules formed through the linking of much smaller, repeating subunits
    c. monomer--the reactant molecule which serves as a building block in polymer synthesis
    d. segmer--the repeating subunit of a polymer
    e. addition polymerization--the formation of a polymer via the addition of monomer units (usually alkenes); the product polymer incorporates all the atoms of the monomer building blocks; no byproducts are produced.
    f. condensation polymerization--the formation of a polymer via a reaction which also results in the elimination of a low molecular weight byproduct (like water)
    g. vulcanization--the reaction of natural rubber with sulfur to produce a polymer in which one long chain is crosslinked to others by sulfur bridges
    h. elastomer--a polymer which will stretch and then return to its original length or shape; a polymer that is characterized by its elasticity
    i. copolymer--a polymer synthesized from more than one monomer; the resulting polymeric chain contains units from two or more monomers
    j. plasticizer--a chemical added to processed polymers in order to improve the flexibility of the final product

2. --

3. Cellulose serves as a starting material for the preparation of all four synthetic products. See section 21.1.

4. Thermosetting polymers have extensively cross-linked structures. Thermoplastic polymers are not cross linked, and the polymer chains are free to slip past one another. Thermosetting polymers set permanently on heating and will not soften on reheating; they are high-melting and rigid. Thermoplastic polymers soften on heating and are generally more flexible and lower melting than thermo- setting products. Section 21.3.

5. a double bond - see table 21.1 for examples

6. Rubber is elastic because the polymer chains are coiled and twisted. The stretching of rubber corresponds to a straightening of the coiled chains. Vulcanization crosslinks the chains. After vulcanized rubber has been stretched, the crosslinks serve to pull the chains back into their original alignment. See section 21.4.

7. Low density polyethylene contains highly branched chains which can not interact effectively with one another; the branches get in the way. High density polyethylene contains long unbranched chains which can line up next to one another to permit strong interactions between chains. Low density polyethylene is low melting, soft and pliable. High density polyethylene is high melting and rigid.

8. Plastics do not decay (or do so very slowly). They take up space which is becoming increasingly valuable in a crowded world. Plastics may produce noxious gases, clog incinerators or simply resist incineration.

9. --

10. It may become possible to find other sources for the carbon usually incorporated in "plastics"--living plants, for example--but almost certainly the cost of processing such materials would be very high.

11. -CH₂-C(CH₃)₂-CH₂-C(CH₃)₂-CH₂-C(CH₃)₂-CH₂-C(CH₃)₂-   [-CH₂-C(CH₃)₂-]ₙ

12. [-O-C(=O)-C₆H₄-C(=O)-O-CH₂-C₆H₁₀-CH₂-]ₙ

13. H₂C=CH-C≡N   acrylonitrile   H₂C=CH-C₆H₅   styrene

Chapter 22

1. a. triose--a sugar molecule containing three carbon atoms; glyceraldehyde is an example
    b. aldose--a sugar molecule containing an aldehyde functional group; glucose and ribose are examples
    c. hexose--a sugar molecule containing six carbon atoms; glucose and fructose are examples
    d. disaccharide--a carbohydrate which yields two mono- saccharide units on hydrolysis; sucrose and maltose are examples
    e. polysaccharide--a carbohydrate which yields many monosaccharide units on hydrolysis; a natural polymer synthesized from monosaccharide monomers; starch and cellulose are examples
    f. aldopentose--a sugar containing five carbon atoms and an aldehyde functional group
    g. ketotetrose--a sugar containing four carbon atoms and a ketone functional group
    h. invert sugar--the mixture of glucose and fructose obtained on hydrolysis of sucrose

2. See structures I and II in section 22.2. The prefix D indicates that a sugar belongs to a family in which the bottom-most secondary hydroxyl group extends to the right of the structural formula. In the L-family, this hydroxyl group points to the left.

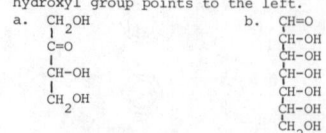

3. a. CH₂OH / C=O / CH-OH / CH₂OH
    b. CH=O / CH-OH / CH-OH / CH-OH / CH-OH / CH₂OH

4. See figure 22.1.
5. See figure 22.3.
6. 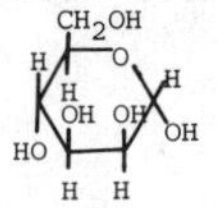 [structure of glucose in cyclic form with CH₂OH, OH, H groups]

7. a,b,c. glucose   d. glucose, galactose
   e. fructose, glucose
8. See section 22.7 for a, b and d.
   c. [structure showing Lactose + H₂O → Galactose + Glucose with H⁺ catalyst]

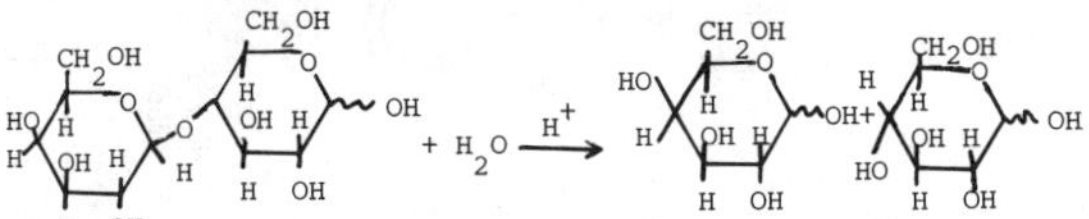

9. a. glucose   b. lactose   c. sucrose
   d. glucose   e. fructose
   f. fructose and glucose   g. sucrose
10. a. α-D-galactose   b. α-D-glucose   c. alpha
    d. Melibiose is a reducing sugar. In the drawing the hemiacetal function is in the alpha arrangement. 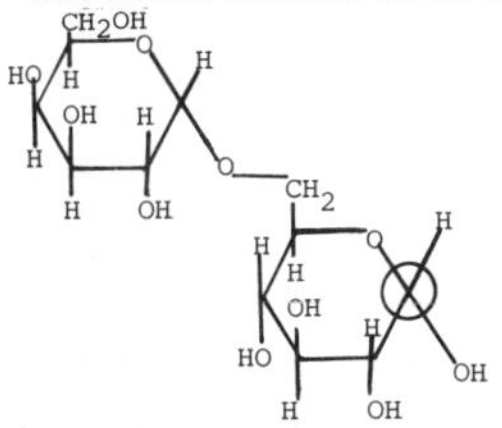
    [structure of melibiose]
11. a. beta
    b. no, the C-1 carbon is fixed as an acetal (not a hemiacetal)
    c. no, no
12. a. Amylose is a polysaccharide in which glucose units are linked in unbranched chains through C-1 and C-4. In amylopectin, in addition to the C-1 to C-4 linked glucose units, there are branches connected through the C-6 alcohol unit. See figure 22.4.
    b. Amylose subunits are connected by alpha acetal links; cellulose by beta.
    c. Glycogen has approximately the same structure as amylopectin. See part a.

Chapter 23

1. a. lipid--a compound isolated from natural sources (plant and animal) which is insoluble in water and soluble in "fat solvents", i.e., nonpolar organic solvents like chloroform and ether. Examples are the triglycerides, prostaglandins, steroids and phosphatides
   b. fatty acid--a carboxylic acid isolated from the hydrolysis of fats; most commonly these acids have long straight chains containing an even number of carbon atoms; examples are lauric, stearic and oleic acids
   c. simple triglyceride--an ester of glycerol and a single type of fatty acid; an example is tristearin
   d. mixed triglyceride--an ester of glycerol and more than one type of fatty acid; examples are: 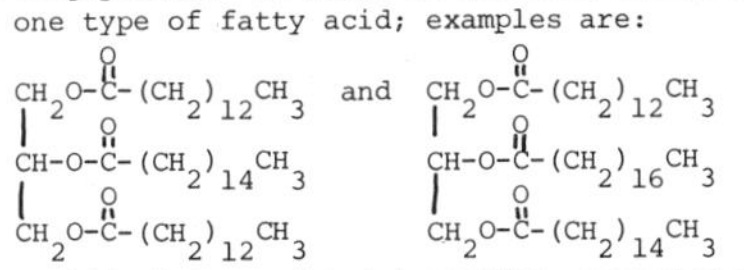
   [two triglyceride structures shown]
   e. fat--a solid mixture of triglycerides; examples are lard and butter
   f. oil--a liquid mixture of triglycerides; examples are corn oil and olive oil
   g. polyunsaturated oil--a mixture of triglycerides which incorporates a high percentage of polyunsaturated fatty acid residues; the polyunsaturated fatty acid content of soybean oil is 70-90%
   h. iodine number--a measure of the degree of unsaturation in a sample of fat or oil; the iodine number equals the number of grams of iodine which will be consumed by 100 grams of fat. The iodine number of butter, a saturated fat ranges between 25 and 40. The iodine number of linseed oil, a polyunsaturated oil, ranges between 170-205.
   i. soap--a water soluble salt of a long chain fatty acid; sodium palmitate is an example-- $CH_3(CH_2)_{14}C(O)O^-Na^+$
   j. wax--the ester of a long chain monohydric alcohol and a long chain fatty acid; hexadecyl stearate is an example-- $CH_3(CH_2)_{15}-O-C(O)-(CH_2)_{16}CH_3$
   k. phospholipid--a lipid which is an ester of phosphoric acid; phosphatides are examples of this class of lipid, [structure shown]
   l. sphingolipids--a lipid which incorporates a sphingosine residue; sphingomyelin is an example
   m. glycolipid--a lipid which contains a sugar residue; cerebrosides are members of this class of lipid
   n. phosphatides--a class of lipid which contains a glycerol residue esterified to two fatty acid residues and one phosphate residue bonded to an aminoalcohol; cephalins and lecithins are examples of this class of lipid
   o. lecithin--a phosphatide containing a choline residue; [structure shown]
   p. cephalin--a phosphatide containing an ethanolamine residue; [structure shown]
   q. cerebroside--a glycolipid incorporating a galactose residue, a sphingosine residue and a fatty acid residue; [structure shown]
   r. steroid--a nonsaponifiable lipid incorporating a characteristic four-ring system; cholesterol is an example 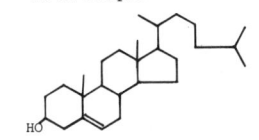
   [cholesterol structure]
   s. prostaglandin--a class of lipids with hormone-like activity which contain a substituted five-carbon ring; see section 23.10 for examples
   t. saponifiable lipid--those lipids which can be hydrolyzed in base; most saponifiable lipids are esters. The fats, waxes and phosphatides are examples of saponifiable lipids.
   u. nonsaponifiable lipid--lipids which are not hydrolyzed in base; steroids are nonsaponifiable lipids
   v. micelle--a spherical arrangement of polar lipid molecules in aqueous solution in which the polar heads are directed outward and the nonpolar tails are directed inward. See figure 23.12a.
   w. monolayer--an arrangement of polar lipid molecules in aqueous solutions in which the lipid molecules are arranged in a single layer at the water's surface with polar heads in the water and the nonpolar tails extended outward from the surface. See figure 23.12b.

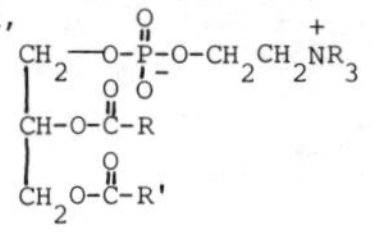

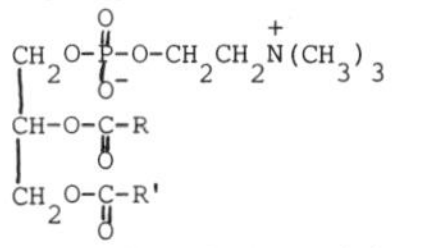

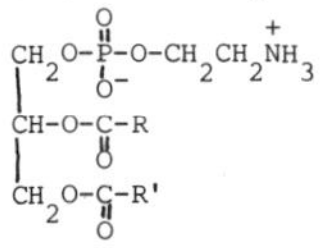

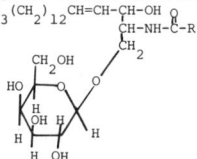

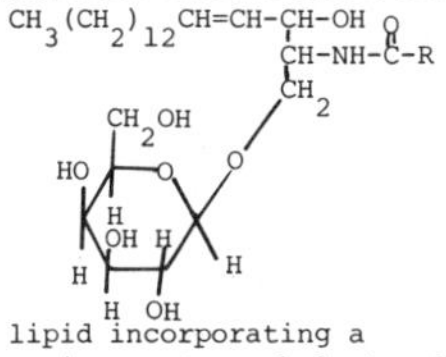

x. bilayer--an arrangement of polar lipid molcules in aqueous solution in which the lipids are arranged in two layers; the nonpolar tails of the lipids are sandwiched in between the polar heads which form the two outer layers of the "sandwich". See figure 23.12c.

2. a.
$$\begin{array}{l} CH_2O-\overset{O}{C}-C_{17}H_{35} \\ CH-O-\overset{O}{C}-C_{17}H_{35} \\ CH_2O-\overset{O}{C}-C_{17}H_{35} \end{array}$$

b. $CH_3(CH_2)_{17}-O-\overset{O}{C}-(CH_2)_{14}CH_3$

c. $CH_3(CH_2)_7CH=CH(CH_2)_7\overset{O}{C}-O^-Na^+$

d. $\left(CH_3-(CH_2)_{12}-\overset{O}{C}-O^-\right)_2 Ca^{2+}$

e. $CH_3(CH_2)_4CH=CHCH_2CH=CH(CH_2)_7\overset{O}{C}-OH$

f. $CH_3CH_2CH=CHCH_2CH=CHCH_2CH=CH(CH_2)_7\overset{O}{C}-OH$

3. Corn oil - Vegetable oils are more highly unsaturated than animal fats. The greater the unsaturation in a molecule, the higher its iodine number is. See section 23.4.
4. Liquid margarine - Highly unsaturated fats tend to have lower melting points than saturated fats. The liquid margarine has the lower melting point, is more unsaturated, and, therefore, has a higher iodine number.
5. Hydrolysis yields fatty acids, and oxidation produces compounds like malonaldehyde. See section 23.4.
6. See section 23.5.
7. Soaps are converted to free fatty acids in acidic solutions:

$C_{17}H_{35}\overset{O}{C}-O^-Na^+ \xrightarrow{H^+} C_{17}H_{35}\overset{O}{C}-OH$

Soaps form insoluble salts in hard water:

$2\ C_{17}H_{35}\overset{O}{C}-O^-Na^+ + Ca^{2+} \longrightarrow (C_{17}H_{35}\overset{O}{C}-O^-)_2\ Ca^{2+} + 2\ Na^+$

soluble   insoluble

8. The calcium, magnesium and iron salts of synthetic detergents are water-soluble. See section 23.5.
9. Tristearin in both cases
10.
$$\begin{array}{l} CH_2O-\overset{O}{C}-(CH_2)_7CH=CH(CH_2)_7CH_3 \\ CH-O-\overset{O}{C}-(CH_2)_7CH=CH(CH_2)_7CH_3 + 3\ NaOH \longrightarrow \\ CH_2O-\overset{O}{C}-(CH_2)_7CH=CH(CH_2)_7CH_3 \end{array}$$

$$\longrightarrow \begin{array}{l} CH_2OH \\ CH-OH \\ CH_2OH \end{array} + 3\ CH_3(CH_2)_7CH=CH(CH_2)_7\overset{O}{C}-O^-Na^+$$

11. a. and c. saponifiable   b. and d. nonsaponifiable
12. a, b, and d. glycerol   f. sphingosine
    c, e, and g. neither
13. Only the phosphatides (d) contain phosphate units.
14.

15. Like soaps, the bile acids are emulsifying agents. They, too, have an ionic and a nonpolar end which permit them to interact with both fats and water. See section 23.9.
16. See section 23.10.

Chapter 24

1. $\overset{+NH_3}{R-\underset{H}{C}-COO^-}$ (sometimes written in nonionized form as $\overset{NH_2}{R-\underset{H}{C}-COOH}$)

2. a. $\overset{+}{H_3N}-CH_2-COO^-$

   b. $\overset{+}{H_3N}-\underset{CH_3}{\overset{CH_3}{CH}}-COO^-$

   c. $\overset{+}{H_3N}-\underset{CH_2-C_6H_5}{CH}-COO^-$

   d. $\overset{+}{H_3N}-CH_2-\overset{O}{C}-NH-\underset{CH_3}{CH}-COO^-$

   e. $\overset{+}{H_3N}-\underset{CH_3}{CH}-\overset{O}{C}-NH-CH_2-COO^-$

   f. $\overset{+}{H_3N}-\underset{CH_2-C_6H_5}{CH}-\overset{O}{C}-NH-CH_2-\overset{O}{C}-NH-\underset{CH_3}{CH}-COO^-$

   g. See table 24.1.   h. See table 24.1.

   i. $\overset{+}{H_3N}-\underset{CH_2-SH}{CH}-COO^-$

   j. $\overset{+}{H_3N}-\underset{\underset{S}{\underset{|}{CH_2}}}{CH}-COO^-$, $\overset{+}{H_3N}-\underset{CH_2-S}{CH}-COO^-$

   k. $H_2N-CH_2-COO^-$

   l. $\overset{+}{H_3N}-\underset{CH_3}{CH}-COOH$

   m. $\overset{+}{H_3N}-\underset{CH_2-C_6H_5}{CH}-COO^-$, $\overset{+}{H_3N}-CH_2-COO^-$, $\overset{+}{H_3N}-\underset{CH_3}{CH}-COO^-$

3. The only difference is the molecular weight. Polypeptides are amino acid polymers with molecular weights below 10 000. Proteins are those with weights above 10 000.
4. a. peptide bond--the amide link which connects one amino acid residue to another in a peptide or protein molecule; $-\overset{O}{C}-NH-$
   b. sequence of amino acids--the order in which amino acid residues are connected to one another in a peptide or a protein; the primary structure of a protein
   c. essential amino acid--an amino acid which must be supplied in the diet because it cannot be synthesized within the body
   d. disulfide linkage--a connection between two peptide chains or two parts of a single peptide chain formed by the reaction of two cysteine units
   e. salt bridge--an interaction between an acidic side chain of one amino acid residue and a basic side chain of another which serves as a bond between two peptide chains or between two parts of the same chain
   f. hydrophobic interaction--the mutual interaction of nonpolar amino acid side chains which removes them as much as possible from contact with the aqueous medium
   g. primary structure--the sequential arrangement of amino acids in a protein which is established and maintained by peptide bonds
   h. secondary structure--a three-dimensional arrangement of proteins with reference to some axis. Two such arrangements are the alpha helix and the pleated sheet.
   i. tertiary structure--the three-dimensional arrangement of proteins which involves the folding of the peptide chain and brings into close proximity amino acid residues which are located relatively far apart along the chain
   j. quaternary structure--the three-dimensional arrangement of the peptide subunits with respect to one another in a protein which contains more than one peptide chain
   k. zwitterion--also called a dipolar ion or an inner salt; a compound which contains an acidic and basic group which interact with one another to form a cation and an anion
   l. isoelectric point--the pH at which the positive and negative charges on an amino acid or a protein just balance; at its isoelectric point, a compound as a whole is neutral
   m. a complete (adequate) protein--a protein which incorporates all of the essential amino acids
   n. globular protein--proteins which form colloidal dispersions in aqueous solutions and whose

tertiary structure gives a roughly spherical shape; albumins and globulins are examples
    o. fibrous protein--proteins which are characteristically stronger and more resistant to denaturation than the globular proteins. Fibrous proteins have a fiber-like structure and are insoluble in aqueous solutions; examples are the collagens of connective tissue and the keratins of hair and nails.
5. See section 24.3.
6. Three problems associated with a strict vegetarian diet are:
    a. Such a diet is more likely to be lacking in some essential amino acid.
    b. Vitamin B$_{12}$ cannot be obtained from plant products.
    c. Calcium and iron and riboflavin and other vitamins may by inadequately supplied.
7. See section 24.9. The secondary structure of silk is termed the <u>pleated sheet</u> conformation.
8. $H_3\overset{+}{N}-CH-\overset{O}{\underset{CH_3}{C}}-NH-CH_2-\overset{O}{C}-O^- + H_2O \longrightarrow H_3\overset{+}{N}-CH-\overset{O}{\underset{CH_3}{C}}-O^- + H_3\overset{+}{N}-CH_2-\overset{O}{C}-O^-$
9. 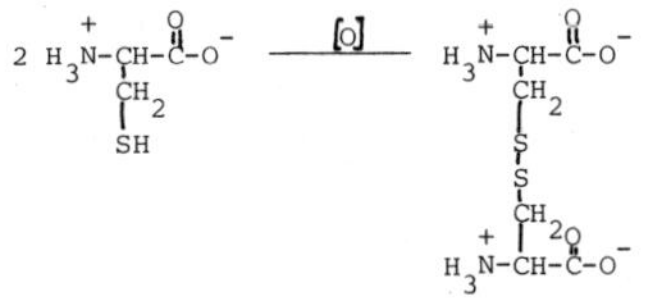 $2\ H_3\overset{+}{N}-CH-\overset{O}{C}-O^- \xrightarrow{[O]} H_3\overset{+}{N}-CH-\overset{O}{C}-O^-$
   (with CH$_2$-SH groups oxidizing to disulfide bridge)
10. See section 24.10. The secondary structure of wool has been given the designation <u>alpha helix</u>.
11. See section 24.11.
12. Wool and silk are fibrous proteins. Myoglobin is a globular protein.
13. Ile·Tyr·Cys-SH
            |
            SH
    Glu·Asp·Cys·Pro·Leu·Gly   or
    Cys·Tyr·Ile·Glu·Asp·Cys·Pro·Leu·Gly

    No. Most evidence indicates that changes in the tertiary structure of a peptide effectively inactivates the molecule.
14. The sequence of amino acid residues in the polypeptide chain may vary from one species to the next.
15. Proteins can be denatured by heat, change of pH, ultraviolet radiation, hydrogen-bonding solvents like ethyl alcohol, alkaloidal reagents like tannic acid, heavy metals like mercury. See section 24.13.
    Denaturation is usually irreversible, but some denatured proteins, under proper conditions, can reestablish their secondary or tertiary structure and regain biological activity.
    Secondary and tertiary structures are usually disrupted when a protein is denatured.

Chapter 25

1. <u>DNA</u>   <u>RNA</u>
   a. adenine and guanine   adenine and guanine (purines)
      (purines)
      cytosine and thymine   cytosine and uracil (pyrimidines)
      (pyrimidines)
   b. deoxyribose   ribose
   c. primary genetic   messenger RNA-transcribes the
      material (genes),   genetic message from DNA and
      carries the genetic   serves as the template for
      blueprint through   protein synthesis
      cell division, directs   transfer RNA-responsible for
      protein synthesis   picking up and transferring
                                       correct amino acid to the
                                       growing peptide chain
   d. double stranded,   single stranded; some portions
      double helix   may have a base-paired double
                                       helix structure which results
                                       when part of a strand folds
                                       back on itself
2. a. adenine   b. guanine and deoxyribose
   c. uracil and ribose   d. cytosine and ribose

3. 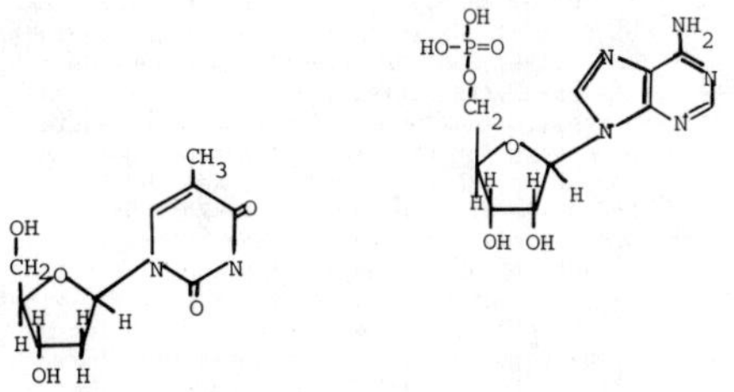 [structures shown]

   nucleoside   nucleotide
   pyrimidine   purine
   incorporated in DNA incorporated in RNA

4. All are derivatives of adenosine and differ only in the number or arrangement of the phosphate units attached. See figure 25.3 and 25.4 for structures.

5. 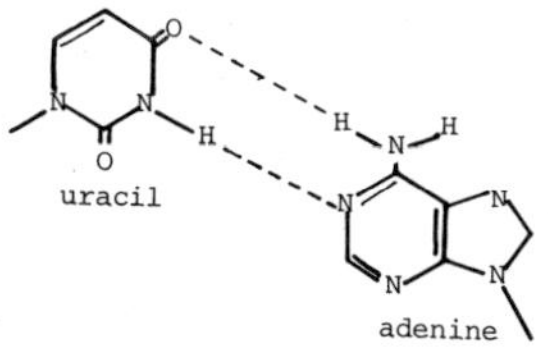 [structure: uracil·····adenine base pair]

6. Pairing cytosine with adenine and thymine with guanine results in fewer interactions (hydrogen bonds) between the members of each pair.

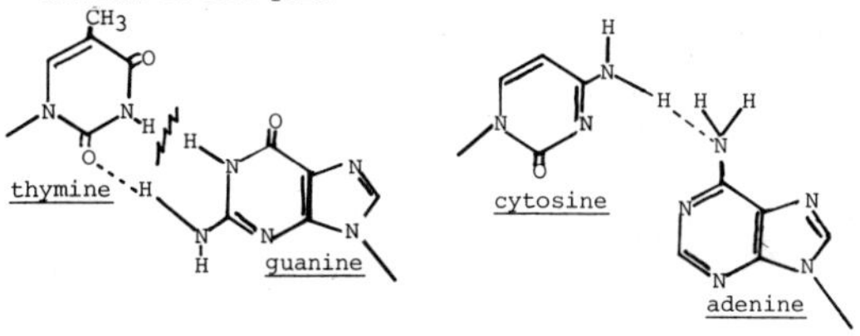

   [structures: thymine···guanine and cytosine···adenine]

7. See figure 25.6.
8. See section 25.6.
9. See section 25.7.
10. There are many possibilities. Here are a few. Each of these things may be responsible for mutations which result from changes in the DNA of a germ cell. A relatively minor change in a DNA molecule may result in an incorrectly formed enzyme which can then be responsible for a metabolic disorder. The cause of the uncontrolled growth associated with cancer cells is still not known. Perhaps some subtle change in DNA structure is responsible.

Chapter 26

1. a. enzyme--a protein which serves as a biological catalyst; examples are lysozyme, urease, trypsin and thrombin
   b. substrate--the compound or type of compound which an enzyme acts upon; the reactant molecule in the reaction catalyzed by an enzyme; the substrate for the enzyme urease is urea
   c. optimum pH--the pH at which an enzyme exhibits its maximum activity; the optimum pH for the enzyme pepsin is 1.6, which means the enzyme is well suited to the acidic fluid of the stomach
   d. optimum temperature--the temperature at which an enzyme exhibits its maximum activity
   e. active site--the portion of an enzyme which comes in contact with the substrate and which is primarily responsible for the enzyme's catalytic effect; it is proposed that the substrate must contain a complementary site whose shape and polarity permit the substrate to bind to the active site of the enzyme as a lock and key fit together
   f. allosteric site--a portion of an enzyme molecule, other than the active site, which will bind an

inhibitor molecule; it is proposed that the binding of the inhibitor changes the overall conformation of the enzyme and thus reduces its catalytic activity
- g. cofactor--a required component, nonprotein in nature, necessary for the proper functioning of an enzyme; metal ions (such as Mg^{2+} and Zn^{2+}) and a number of vitamins serve as cofactors
- h. coenzyme--organic molecules which serve as enzyme cofactors; all coenzymes are nonprotein and many are vitamins or derivatives of vitamins
- i. proenzyme--an inactive form of an enzyme which can be converted in the body to the active form (frequently by hydrolysis and removal of a peptide unit); pepsinogen and prothrombin are proenzymes which, on activation, yield the enzymes pepsin and thrombin respectively
- j. feedback inhibition--a regulatory mechanism in which an enzyme is inhibited by the end product of the reaction it catalyzes. See section 26.6.
- k. apoenzyme--the protein portion of an enzyme which requires a coenzyme to carry out its catalytic function
2. See section 26.2.
3. Enzymes act by lowering the activation energy for a given reaction. They do not so much cause a reaction to take place as increase the rate at which the reaction takes place. A reaction may proceed so slowly that changes are not readily observable unless an enzyme is present and increases the rate significantly (sometimes by a factor of a million). Thus the enzyme may appear to cause a reaction where there was none before. See section 26.4.
4. Enzymes do not change the extent of a reaction; they merely change the speed with which the equilibrium concentrations can be reached. Enzymes catalyze the forward and the reverse reaction. See section 26.4.
5. All enzymes are proteins. Some require metal ions or organic cofactors to carry out their catalytic functions.
6. See section 26.4 and figure 26.4.
7. a. absolute specificity--the enzyme's catalytic activity is restricted to the reaction of a single specific substrate; urease is an example of an enzyme which exhibits absolute specificity
- b. stereospecificity--the enzyme's activity is restricted to the reaction of one member of a pair of enantiomers; arginase is an example of a stereospecific enzyme since it catalyzes the hydrolysis of L-arginine, but not D-arginine
- c. reaction specificity--the enzyme's activity is restricted to reactions of a specific type, e.g., hydrolysis
- d. linkage specificity--the enzyme's activity is restricted to a specific chemical bond; thrombin is a linkage specific enzyme which cleaves peptide bonds connecting arginine to glycine

See section 26.3.
8. See section 26.9.
- a. heat--acts by denaturing the protein molecule; the energy absorbed by the enzyme disrupts secondary, tertiary and quaternary structure necessary for catalytic function
- b. lead ions--denature proteins by reacting with sulfhydryl groups on the enzyme; sulfhydryl groups often appear at the active site of an enzyme and are required in free form for the functioning of the enzyme; in addition, the lead-protein combination usually coagulates
- c. ozone--ozone is an oxidizing agent and can oxidize a number of critical groups (sulfhydryl groups, for example) on the enzyme molecule; ozone is a sufficiently powerful reagent that many bonds within the enzyme could be broken
- d. excess hydrogen ion--most enzymes achieve maximum activity within a fairly narrow range of pH; increasing the acidity of the medium will ultimately produce a pH below optimum; acids disrupt salt bridges, hydrogen bonds, etc., and denature proteins, bringing about coagulation
9. Over a limited range the rate of an enzyme catalyzed reaction increases with increasing substrate concentration. At the point at which the substrate concentration results in the saturation of all available enzyme molecules, the rate levels off and shows no appreciable change with increasing substrate concentration. See section 26.5.
10. EDTA chelates the lead ions and this complex can be excreted from the body. See section 26.9

11. Cyanide reacts with cytochrome oxidases by tying up the electrons which these enzymes normally supply in reduction reactions. The deactivation of the oxidases prevents reduction of oxygen in the cell, which means that cell respiration is interrupted. See section 26.9.
12. Thiosulfate converts the cyanide ion to the relatively nontoxic thiocyanate ion. See section 26.9.
13. Some biochemical reactions must be initiated on very short notice. In some cases, these reactions may be beneficial to the organism only under certain specific conditions and harmful or fatal otherwise. Proenzymes provide a means of controlling such reactions. The proenzyme can be activated on very short notice and can then catalyze the necessary reaction. But until conditions trigger the activation of the proenzyme, no enzyme is present to catalyze the reactions. See sections 26.7 and 26.10.
14. Enzymes are used in laundry preparations, as meat tenderizers, and to "predigest" protein in baby foods. There are also a number of medical applications of enzymes. See section 26.10.

Chapter 27

1. a. vitamin--an organic compound required in minute amounts for the proper functioning of an organism but not synthesized by the organism; if its diet lacks a particular vitamin, the organism exhibits the symptoms of a specific vitamin-deficiency disease; see sections 27.2 through 27.7 and table 27.1 for examples
- b. hormone--a compound produced in trace amounts by the endocrine glands which triggers biochemical and physiological responses in a specific target organ; examples are given in table 27.3
- c. provitamin--a vitamin precursor; a compound which is converted within the body to an active vitamin; β-carotene is a provitamin which yields two molecules of vitamin A when cleaved within the body. See section 27.2.
- d. androgen--a steroid which serves as a male sex hormone; testosterone is an example
- e. estrogen--a steroid which serves as a female sex hormone; estradiol and estrone are examples
2. Both vitamins and hormones are organic molecules which produce physiological effects when present in trace amounts. They differ in that vitamins cannot be synthesized by the body and are required in the diet, whereas hormones are synthesized by the endocrine glands. Vitamins and hormones also perform different biochemical functions.
3.

Fat Soluble Vitamins	Food Source	Deficiency Disease
Vitamin A	fish liver oils, liver, eggs, fish, butter; carrots tomatoes, green vegetables supply the provitamin	night blindness, xerophthalmia, arrested growth
Vitamin D	fish liver oils, fortified milk	rickets, osteomalacia
Vitamin E	wheat germ oil, green vegetables, vegetable oils, egg yolk and meat	muscular weakness, sterility, scaly skin
Vitamin K	green leafy vegetables	increased blood clotting time

See table 27.1 for information on the water-soluble B vitamins. Vitamin C is also a water soluble vitamin. Good food sources for vitamin C are citrus fruits, tomatoes, green peppers and strawberries. Its deficiency disease is scurvy.
4. Many vitamins are modified in the body to produce active coenzymes. See section 27.6 and table 27.2.
5. Synthetic and natural vitamin C are identical. All vitamin C is ascorbic acid.
6. Water-soluble vitamins would be lost. They would be extracted by the water used in "boiling" vegetables and then lost when this water is poured off.
7. Such a vitamin pill might satisfy one's need for fat-soluble vitamins, which can be stored in the body. However, the water-soluble vitamins cannot be stored, and excesses over the immediate requirement are excreted. Thus, even if a month's supply of the vitamins were

present in the pill, most would be eliminated from the body within a few days. A person depending on the once-a-month vitamin pill would lack water-soluble vitamins for most of the month.
8. See section 27.2.
9. Vitamin D is called the "sunshine vitamin" because one form of the vitamin is synthesized in the skin of animals through the action of sunlight on 7-dehydrocholesterol. (Section 27.3)
10. Vitamin A and polyunsaturated fatty acids are believed to be protected from oxidation by the presence of vitamin E. (Section 27.4)
11. The "B complex" is the group of water-soluble vitamins commonly found together in a variety of food sources. They are not closely related structurally. See section 27.6 and table 27.1.
12. See section 27.8.
13. See section 27.11.
14. The presence of an ethynyl group (—C≡CH) on the steroid molecule rendered the birth control pill an oral contraceptive.
15. The Delaney clause of the Food and Drug Act of 1958 requires that any agent shown to cause cancer in humans or laboratory animals not be used as a food additive. The clause does not apply to drugs. As soon as evidence was obtained that DES was a carcinogen, it was banned as a food additive. See section 27.12.

Chapter 28

1. See section 28.1.
   Functions of the blood include transport of O_2 from lungs to tissues, transport of CO_2 from tissues to lungs, transport of nutrient from intestines to cells, transport of metabolic wastes from cells to excretory organs, and transport of hormones from endocrine glands to target tissues. The blood also carries white blood cells and immunoglobulins which represent a defense mechanism of the body, helps maintain body temperature, helps maintain acid-base balance, and helps maintain fluid balance.
2. plasma--after addition of an anticoagulant, the liquid portion of the blood (the plasma) is separated from the formed elements
   serum--no anticoagulant is added and the blood is allowed to clot; the clot and formed elements are separated, leaving a liquid portion (serum)
   See section 28.1.
3. Calcium ion is necessary for the clotting of the blood.
4. The formed elements in the blood are the organized structural units (cells) suspended in the blood plasma. They include the red blood cells, the white blood cells and the platelets.
5. erythrocyte--red blood cell; contains hemoglobin and is responsible for oxygen and carbon dioxide transport
   leukocyte--white blood cell; part of the body's defense mechanism; attacks, immobilizes and destroys infectious bacteria
   thrombocyte--platelet; involved in the clotting mechanism of the blood
   See section 28.1.
6. Section 28.3. Four protein fractions of the blood include albumins, globulins, fibrinogen, and prothrombin.
7. Positive ions: Na^+, K^+, Ca^{2+}, Mg^{2+}
   Negative ions: HPO_4^{2-}, Cl^-, HCO_3^-, SO_4^{2-}
   See section 28.2 and table 28.1.
8. a. $Cl^- = 120$ meq/ℓ
      $Ca^{2+} = 1.25$ meq/ℓ
      $Na^+ = 140$ meq/ℓ
   b. Chloride and sodium ions fall within the normal ranges, but calcium ion is significantly below the normal range. This condition accompanies some forms of kidney disease, vitamin D deficiency, an underactive parathyroid.
9. Interstitial fluid is the fluid surrounding cells in body tissue. This fluid, once it is taken into the lymphatic system, is called lymph.
10. See section 28.3.
11. Edema refers to the abnormal accumulation of fluid in body tissues. It can result from an imbalance in the osmotic pressures of the blood and interstitial fluid. See section 28.3.
12. Shock is a medical condition characterized by a sudden drop in blood pressure and a resulting decrease in the oxygen transporting capabilities of the blood. See section 28.3.

13. The main blood buffers are $H_2PO_4^-/HPO_4^{2-}$, H_2CO_3/HCO_3^-, and the plasma proteins.
14. See section 28.5.
15. respiratory acidosis--inadequate ventilation caused, for example, by a disease of the lungs (like emphysema)
    respiratory alkalosis--hyperventilation
    metabolic acidosis--buildup of acidic metabolic products like lactic acid, ketoacids or inorganic acids
    Section 28.5.
16. See section 28.5.
17. When bicarbonate ion diffuses from red blood cells, chloride ions diffuse into these cells to maintain electrical neutrality. See section 28.5.
18. Antibodies (immunoglobulins) are specialized proteins synthesized in response to an invading, non-native macromolecule. They are specific for each foreign molecule and react with and incapacitate it. Antigens are those substances which trigger the formation of antibodies. Section 28.4.
19. See section 28.4.
20. Antibodies are found in the gamma globulin fraction of blood proteins. They are formed in the lymph nodes.
21. A vaccine is a weakened form of a pathogenic microorganism used to trigger the immune response. Its purpose is to prime the body's defense mechanism for a more effective response to attack by a more virulent form of the microorganism. Section 28.4.
22. Both involve activation of the immune response. Immunity refers to the presence of antibodies specific to some pathogenic microorganism (such as the polio virus). Tissue rejection refers to the presence of antibodies specific to transplanted foreign tissue (such as a kidney transplant). Section 28.4.
23. Perspiration cools the body through evaporation from skin surfaces; the heat of vaporization is obtained from the body. Section 28.9.
24. Section 28.6.
25. Bilirubin is a breakdown product of hemoglobin; it is formed from the heme units of the hemoglobin molecule.
26. The lymphatic system:
    1. returns components of interstitial fluid to the blood circulation
    2. carries fats from the intestines to the blood circulation
    3. is the site of white blood cell and antibody synthesis
    4. serves as a filtering system to remove dead cells and bacteria from circulation
    Section 28.7.
27. See section 28.8.
28. 1. liquid intake
    2. body temperature
    3. amount of perspiration
    4. other liquid losses (e.g., because of diarrhea)
    Section 28.8.
29. urea, uric acid, breakdown products of bilirubin, and, in certain pathological conditions, glucose (blood sugar) among others
30. sensible perspiration--perspiration produced by the sweat glands
    insensible perspiration--perspiration not produced via the sweat glands; perspiration lost directly through the skin or from the respiratory tract
31. organic--lipids, urea, creatinine, lactic acid, pyruvic acid, some drugs
    inorganic--electrolytes (e.g., Na^+, Cl^-, Ca^{2+}), water
    Section 28.9.
32. inner layer--mucous
    middle layer--lacrimal secretions (water solution)
    outer layer--nonvolatile oily protective layer
    Section 28.10.
33. Lysozyme is a bacteriocide. It is an enzyme which catalyzes the rupture of bacterial cell walls. Section 28.10.
34. Cow's milk is a perfect food for calves. It is a nutritious food for humans and is a source of complete protein, but must be supplemented by other foods in the human diet.
35. Nature has arranged for mother's milk to be tailored to the nutritional needs of the offspring (e.g., mammals which live in cold climates are supplied with higher levels of energy rich fats). In some cases, the mother's immunoglobulins may be transferred in the milk to the child, conferring on the child a temporary passive immunity. Section 28.11.

Chapter 29

1.

Enzyme	Location of Action	Function
a. salivary amylase (ptyalin)	saliva (mouth)	cleaves alpha-acetal linkages in starch
b. pepsin	stomach	hydrolyzes protein at tyrosine or phenylalanine linkages
c. trypsin	small intestine	hydrolyzes protein at arginine or lysine linkages
d. chymotrypsin	small intestine	hydrolyzes protein at tyrosine or phenylalanine linkages
e. pancreatic amylase	small intestine	completes hydrolysis of starch to maltose and (to a small extent) to glucose
f. sucrase	small intestine	hydrolyzes sucrose to glucose and fructose
g. lactase	small intestine	hydrolyzes lactose to glucose and galactose
h. pancreatic lipase (steapsin)	small intestine	catalyzes the hydrolysis of fats to monoglycerides and fatty acids
i. carboxypeptidase	small intestine	splits off the amino acid at the free carboxyl end of a protein
j. dipeptidase	small intestine	hydrolyzes dipeptides
k. enterokinase	small intestine	converts trypsinogen (a proenzyme) to trypsin (a protease)
l. nucleotidase	small intestine	catalyzes the hydrolysis of nucleotides to nucleosides and phosphoric acid
m. maltase	small intestine	hydrolyzes maltose to glucose

2. Mucin is a glycoprotein (a molecule which contains a protein portion and a carbohydrate portion) found in the saliva. When food is chewed it is coated with a slick layer of mucin which lubricates the food, making it easier to swallow. Section 29.2.

3. Pepsinogen is a proenzyme found in the stomach. It is converted to the active protease pepsin by the action of hydrochloric acid present in the stomach. This conversion is autocatalyzed, i.e., as pepsin is formed it catalyzes the formation of more pepsin. Section 29.3.

4. "Intrinsic factor" is a mucoprotein (a class of glycoproteins, see answer for question 2) produced by the stomach lining and required for the absorption from the diet of vitamin B_{12}. Section 29.3.

5. The bicarbonate ion (HCO_3^-) present in pancreatic juice makes this fluid slightly alkaline because of the following equilibrium:

   $$HCO_3^- + H_2O \rightleftharpoons H_2CO_3 + OH^-$$

   In essence, pancreatic juice can be considered a solution of sodium bicarbonate, the salt of a weak acid and strong base. Such a solution is expected to be slightly basic.

6. A pH of 8 means the enzyme is most effective in slightly basic media; therefore, the enzyme is more suited to the somewhat alkaline solutions of the small intestine, rather than the strongly acidic chyme of the stomach.

7. On entering the duodenum, the acidic chyme is neutralized by the alkaline pancreatic juice. The reaction is given in the equation:

   $$HCl + NaHCO_3 \longrightarrow NaCl + H_2O + CO_2$$

   or

   $$H^+ + HCO_3^- \longrightarrow H_2CO_3 \longrightarrow H_2O + CO_2$$

8. Vomiting empties the contents of the stomach, which means that considerable chloride ion is lost because the stomach juices are rich in HCl. The body attempts to compensate for the loss by reestablishing a level of HCl in the stomach. The chloride ion required is obtained from other body fluids.

9. The bile salts are steroid molecules containing ionic groups. The nonpolar portion of the steroid interacts strongly with the nonpolar fats. The ionic portion interacts strongly with the aqueous digestive juices. Large fat globules are broken down into smaller droplets. As the nonpolar portions of the bile salts bury themselves within the fat droplet, the polar portions form a charged layer at the surface of the droplet. The like-charged surface layers prevent the droplets from coalescing with one another and, at the same time, permit a thorough mixing of the droplets and the aqueous medium. This emulsifying action permits the enzymes in the aqueous medium far greater contact with the fats to be hydrolyzed. Section 29.6.

10. Gallstones are compact, solid deposits (sometimes called concretions) formed within the gallbladder. They sometimes move to the duct leading from the liver to the small intestine, where they may block the flow of bile. The chief constituent of gallstones is usually cholesterol.

11.
    a. monosaccharides including glucose, fructose and galactose
    b. monoglycerides and fatty acids
    c. amino acids

12. Fats are absorbed through the villi and into the lymphatic system rather than directly into the blood.

13.
    a. gastric juices--flow initiated by the protein hormone gastrin, which is produced in the stomach
    b. pancreatic juice--flow initiated by the protein hormone secretin, which is produced in the intestine and carried to the pancreas through the circulatory system
    c. bile--the release of bile stored in the gallbladder is signalled by the protein hormone cholecystokinin

14. The feces contain undigested carbohydrates (particularly cellulose), proteins (particularly those of connective tissue), fats, dead mucosal cells, remnants of digestive fluids, bacteria, and are normally colored by the breakdown products of heme. Section 29.8.

15. Black, tarry stool--bleeding in the upper digestive tract; color due to the presence of methemoglobin
    Clay-colored stool--blockage of bile flow (gallstones?); color due to the absence of breakdown products of heme

16. See section 29.7.

Chapter 30

1.
   a. blood sugar--glucose or dextrose
   b. hypoglycemia--a condition characterized by an abnormally low blood sugar level
   c. hyperglycemia--a condition characterized by an abnormally high blood sugar level
   d. renal threshold--the minimum concentration of a substance in the blood which will result in the transfer of the substance to the urine; when the renal threshold for a particular compound is exceeded, that compound begins to show up in the urine
   e. glycolysis--the anaerobic energy-producing breakdown of glucose to lactic acid
   f. glycogenolysis--the breakdown of the polysaccharide glycogen to glucose
   g. glycogenesis--the conversion of the monosaccharide glucose to the polysaccharide glycogen
   h. glucose tolerance test--a standard procedure designed to evaluate an individual's ability to regulate blood sugar levels; the test is used to diagnose diabetes mellitus
   i. galactosemia--a potentially fatal hereditary disease caused by an inability to synthesize the enzyme responsible for converting galactose to glucose; affected infants cannot tolerate milk

2.
   a. insulin--regulates the absorption of blood glucose by tissues; results in a decrease in blood glucose levels
   b. glucagon--regulates breakdown of glycogen to glucose; results in an increase in blood glucose levels
   c. epinephrine--increases breakdown of glycogen to glucose; results in an increase in blood glucose levels
   d. cortisone and cortisol--stimulate conversion of amino acids to glucose; results in an increase in blood glucose levels
   e. human growth hormone--stimulates glucose production in the liver and glucagon secretion in the pancreas; results in an increase in blood glucose levels

   Section 30.2.

3. No. Two examples of nonpathological conditions which will cause glucose to appear in the urine can be cited. The renal threshold for glucose may be exceeded shortly after a meal rich in carbohydrates is consumed; also, because of the action of epinephrine, a frightened individual may exhibit glucosuria.

4. Insulin is a peptide and will be subjected to and deactivated by the same digestive enzymes which hydrolyze protein foods.
5. The oral drugs are nonprotein and act to stimulate the production of insulin by the pancreas. Section 30.2.
6. The storage form of carbohydrate is glycogen, found in muscle and the liver. Glycogen is a polymer formed by the coupling of glucose units through 1,4-α-acetal linkages, with occasional branches involving 1,6-α-acetal linkages. See section 30.3.
7. Glycogen synthetase is responsible for forming the 1,4-linkages, and the branching enzyme forms the 1,6-linkages in glycogen. Section 30.3.
8. lactic acid
9. Step 6 in the Embden-Meyerhof pathway involves the oxidation of glyceraldehyde-3-phosphate to 1,3-diphosphoglyceric acid. The oxidizing agent is NAD^+. Section 30.4.
10. These high energy compounds transfer their phosphate units to ADP to form ATP, which serves as the energy currency of the cell. Section 30.4.
11. See section 30.4.
    In glycolysis, pyruvic acid is converted to lactic acid. In fermentation, pyruvic acid is converted first to acetaldehyde, then to ethyl alcohol.
12. The Krebs Cycle provides energy to the cell in the form of energy-rich ATP molecules. Section 30.5.
13. as two CO_2 molecules
14. NAD^+ and FAD
15. The electron transport system is responsible for transferring the electrons originally obtained from acetyl CoA in the Krebs Cycle to oxygen molecules. Another way of looking at it is to regard the respiratory chain as the supplier of all the oxidizing agents (NAD^+ and FAD) required by the Krebs Cycle. The ultimate oxidizing agent in this system is oxygen (O_2). Section 30.6.
16. Cytochromes are globular iron-containing proteins. Most contain one or two heme units. The cytochromes are electron carriers in the respiratory chain and are thus involved in the transfer of electrons obtained from the oxidation of carbohydrates to oxygen molecules.
    Carbon monoxide and cyanide ion bond strongly to the iron of the cytochromes; the complexed cytochromes are unable to participate effectively in the respiratory chain. Section 30.6.
17. The aerobic Krebs Cycle is much more efficient than the anaerobic Embden-Meyerhof pathway.
18. a. ATP--immediate source of energy for muscle contraction
    b. creatine phosphate--provides a means for quickly regenerating ATP by the rephosphorylation of ADP, thus serves as a readily available energy reserve
    c. actin--one component of the structural protein of muscle
    d. myosin--one component of the structural protein of muscle; also the enzyme which catalyzes the removal of phosphate from ATP to supply energy for muscle contraction
    e. actomysin--the protein complex which undergoes the physical contraction or extension associated with muscle action
    f. aerobic pathway--the source of energy for the resting muscle
    g. anaerobic pathway--the source of energy for vigorously exercising muscle; the oxygen debt incurred during anaerobic energy production must ultimately be repaid via the aerobic pathway
    Section 30.7.

Chapter 31

1. a. adipose tissue--the specialized tissue in which fat droplets are stored within cells held together by connective tissue
   b. fat depot--a fat storage site, e.g., the adipose tissue located under the skin; such tissue is in a dynamic state with fats being added to and withdrawn from storage
   c. beta oxidation--a step in the metabolism of a fatty acid in which the carbon atom beta to the carboxyl group of the fatty acid undergoes oxidation; the following step in the fatty acid cycle:
   $$RCHCH_2CSCoA + NAD^+ \rightarrow RCCH_2CSCoA + NADH + H^+;$$
   (with OH and O groups shown)
   occasionally, beta oxidation is used to designate the entire process referred to in this chapter as "the fatty acid spiral"
   d. ketone bodies--acetone, acetoacetic acid and β-hydroxybutyric acid, compounds which are intermediates in the fatty acid cycle or compounds derived from these; abnormally high levels of the ketone bodies lead to the condition known as ketosis (Section 31.7)
   e. ketosis--the condition characterized by abnormally high levels of the ketone bodies in the blood (Section 31.7)
   f. acidosis--the condition which results when the buildup of ketone bodies (two of which are carboxylic acids) causes a significant drop in the pH of the blood; see Section 31.8)
2. Lipids and fatty acids are ordinarily complexed with water-soluble proteins to form lipoproteins. The lipoproteins are readily transported by the blood.
3. acetyl CoA
4. 1. Acetyl CoA serves as fuel for the Krebs tricarboxylic acid cycle.
   2. It serves as a starting material for fatty acid synthesis.
   3. It serves as a starting material for steroid synthesis.
5. The glycerol can enter the glycolysis pathway (Embden-Meyerhof pathway) after being phosphorylated and then oxidized to dihydroxyacetone phosphate. Section 31.3.
6. fatty acid metabolism--dehydrogenation (oxidation) to form a double bond; hydration of the double bond; oxidation of an alcohol to a ketone; cleavage of the β-ketoacid
   fatty acid synthesis--condensation to form a β-ketoacid; reduction of a ketone to an alcohol; dehydration to form a double bond; hydrogenation (reduction) to form the saturated acid
7. Most fatty acids are synthesized by a route which adds a two-carbon unit on each turn of the cycle. Section 31.4. Malonyl CoA is the 3-carbon compound condensed with the growing acid chain. A carbon dioxide unit is lost during the condensation, leaving only two carbon atoms of the original malonyl group attached to the growing chain.
8. Section 31.2.
   1. Adipose tissue acts as insulation to prevent the loss of body heat.
   2. It acts as padding around vital organs to absorb shocks and minimize damage to the organs.
   3. It acts as a storehouse for energy reserves.
9. Overeating; eating more food than is required for replacement of tissue and for energy needs
10. Cytidine triphosphate acts as a carrier molecule. It combines with one reactant molecule and activates it for transfer to a second reactant. See section 31.6 for examples.
11. cholesterol
12.

Glycogen	Adipose Tissue
stored in very limited quantities	can be stored in essentially unlimited quantities
activated as soon as blood sugar levels are depleted	called on after glycogen reserves are depleted
yields about 4 kcal/g; a comparatively inefficient mode of energy storage	yields about 9 kcal/g; a comparatively efficient mode of energy storage

13. Glycogen reserves are used first.
    Fat reserves supply the major part of the body's needs. Proteins, including ultimately structural protein such as that of muscle, supply energy after the fat reserves are exhausted.
14. Section 31.8.
    Two of the ketone bodies are carboxylic acids. A buildup of the ketone bodies produces ketosis first. Eventually the concentration of ketone bodies reaches a level which can no longer be handled by the blood buffers. When the blood buffers are overwhelmed, the pH drops because of the increasing concentration of acids, and acidosis occurs.
15. In starvation, the body shifts to fatty acid metabolism almost exclusively. The resulting production of acetyl CoA may exceed the capacity of the Krebs Cycle to metabolize the acetyl CoA. As the acetyl CoA begins to accumulate, the liver begins to convert it to the ketone bodies. The accumulation of the acidic ketone bodies leads to acidosis. Section 31.8.
16. Lack of insulin prevents glucose from getting into cells from the blood. Tissues switch to fatty acid metabolism for their energy requirements, which leads to the production of acetyl CoA and a resulting production of ketone bodies, leading ultimately to acidosis. The conversion of amino acids to glucose (glyconeogenesis) is also accompanied by the accumulation of ketone bodies. Such gluconeogenesis also takes place when lack of insulin prevents glucose from getting from the blood into cells. Section 31.8.
17. See figure 31.9.
18. Palmitic acid contains 16 carbon atoms and requires 7 turns of the fatty acid spiral to be converted completely to acetyl CoA (eight molecules) at a cost of one ATP

molecule. During the oxidation, 7 FADH$_2$ and NADH molecules are formed. Each FADH$_2$ provides 2 ATP molecules when processed by the respiratory chain. Each NADH molecule yields 3 ATP molecules by the same route. Each acetyl CoA molecule yields 12 ATP molecules when processed through the Krebs Cycle. Thus:

```
        8 acetyl CoA  ─────────→   96 ATP
        7 FADH₂       ─────────→   14 ATP
        7 NADH        ─────────→   21 ATP
                                  ──────
                                  131 ATP
    Less ATP required for oxidation  -1 ATP
                                  ──────
                                  130 ATP
```

19. The lipid storage diseases are hereditary in origin. The coding for enzymes critical to the metabolism of lipids is faulty and results in inactive or nonexistent enzymes. Nerve and brain tissue, which are particularly rich in the glycolipids, exhibit the effects of lipid storage diseases.

Chapter 32

1. a. anabolism--the metabolic process in which molecules required by cells are built up from smaller molecules; a more restricted definition is the buildup of proteins for growth, replacement and repair
    b. catabolism--the metabolic process in which the large molecules of the cell are broken down or decomposed; a more restricted definition refers only to the breakdown of proteins for energy
    c. protein turnover time--the average residence time for a protein molecule in tissue, usually measured as a half-life
    d. positive nitrogen balance--the condition in which more nitrogen is being taken into the body than is being excreted
    e. negative nitrogen balance--the condition in which more nitrogen is being excreted from the body than is being ingested
2. The amino acid pool is a circulating accumulation of amino acids in the blood. The pool is constantly being depleted and renewed. See figure 32.1. Both glycogen and fats are stored in specific, more or less localized areas of the body--glycogen in the liver and muscles, fats in adipose tissue around organs and under the skin. The amino acid "pool" is the blood stream. The amino acids in the pool are simply dissolved in the blood and brought in contact with all body tissue by way of the blood circulation.
3. Contributing to the amino acid pool:
   1. breakdown of dietary protein
   2. breakdown of tissue protein
   3. synthesis of amino acids

Depleting the amino acid pool:
1. synthesis of proteins
2. synthesis of nitrogen-containing nonprotein molecules
3. catabolism of amino acids to produce energy
See section 32.1.

4. a. The person goes into negative nitrogen balance.
    b. The person goes into negative nitrogen balance.
    c. The person goes into positive nitrogen balance.
See section 32.2.

5. Transamination:

$$\begin{array}{c}COOH\\|\\CHNH_2\\|\\R\end{array} + \begin{array}{c}COOH\\|\\C=O\\|\\(CH_2)_2\\|\\COOH\end{array} \longrightarrow \begin{array}{c}COOH\\|\\C=O\\|\\R\end{array} + \begin{array}{c}COOH\\|\\CHNH_2\\|\\(CH_2)_2\\|\\COOH\end{array}$$

α-ketoglutaric acid   glutamic acid

Oxidative Deamination:

$$\begin{array}{c}COOH\\|\\CHNH_2\\|\\(CH_2)_2\\|\\COOH\end{array} \longrightarrow \begin{array}{c}COOH\\|\\C=O\\|\\(CH_2)_2\\|\\COOH\end{array} + NH_3$$

glutamic acid   α-ketoglutaric acid

6. a. aspartic acid
    b. alanine
    c. phenylalanine
7. Transamination reactions yield ketoacids, some of which can be converted to acetyl CoA (and, therefore, to fatty acids); some can be converted to intermediates of the Krebs Tricaroxylic Acid Cycle (and, therefore, to carbohydrates). Section 32.3 and figure 32.9.
8. Most is fed into the urea cycle where it is converted to, then eliminated from the body as, urea. Section 32.5 and figure 32.5.
9. The end product of purine metabolism is uric acid. The ultimate end products of pyrimidine metabolism may be CO_2, H_2O and urea. Cytosine and uracil, however, are first converted to β-alanine, and thymine is first converted to β-aminoisobutyric acid. See section 32.6.
10. No. Both types of bases can be synthesized from available starting materials within the body. Section 32.6.
11. DNA contains the base thymine, which is converted to β-aminoisobutyric acid. In RNA, thymine is replaced by uracil, which is not converted to β-aminobutyric acid.
12. Section 32.4, phenylketonuria (PKU) and albinism. In PKU the inability to convert phenylalanine to tyrosine results in severe mental retardation. In albinism the inability to convert tyrosine to the melanin pigments results in a lack of pigmentation in skin, hair and eyes.
13. A sudden increase in the levels of both SGOT and CPK in the blood indicates relatively severe muscle deterioration and may be an indication of a myocardial infarction (one type of "heart attack"). Section 32.4.